Instructor's Resource Manual

BASIC CONCEPTS OF CHEMISTRY

Instructor's Resource Manual

BASIC CONCEPTS OF CHEMISTRY

Fifth Edition

Alan Sherman

Sharon J. Sherman

Leonard Russikoff

HOUGHTON MIFFLIN COMPANY BOSTON TORONTO
Dallas Geneva, Illinois Palo Alto Princeton, New Jersey

Sponsoring Editor: Richard Stratton
Project Editor: Susan Lee-Belhocine
Assistant Design Manager: Karen Rappaport
Production Coordinator: Frances Sharperson
Manufacturing Coordinator: Sharon Pearson
Marketing Manager: Michael Ginley

Printed in the U.S.A.

ISBN: 0-395-60622-5

ABCDEFGHIJ-B-954322

CONTENTS

PREFACE

This instructor's manual is a supplement to our textbook Basic Concepts of Chemistry, Fifth Edition, and to our laboratory manual Laboratory Experiments for Basic Chemistry, Fifth Edition. These books complement one another by design, but each can be used independently or in conjunction with other books. To provide flexibility, we have avoided explicit cross-references from the textbook to the laboratory manual.

To further complement the textbook, a Study Guide by James R. Braun is also available. In addition, a Solutions Manual containing worked-out solutions to all problems in the text is available for this edition. A set of 52 Transparencies is provided to instructors upon adoption of the text. Finally, a Computerized Test Generator containing items from Parts I, II, and III of this manual is offered free of charge to adopters. This computer ancillary is available for the IBM®-PC and compatibles only.

Part I of this instructor's resource manual contains a unique and useful teaching aid—a mathematics pretest/retest. This section, which can be administered at the start of the course and again at the end, serves a dual purpose. Given as a pretest, it enables students to identify any mathematical weaknesses they might have and gives you, the instructor, insight into their various needs. It also provides a baseline with which you can compare the results of students' later performances over the same material. The test can be administered at the end of the course (or after students have completed work on Supplement A in the textbook, "Basic Mathematics for Chemistry") to build students' confidence or to point out need for further drill and practice with problems involving chemical mathematics. Students should be reassured that the scores on the mathematics pretest/retest are not counted into their grades in the course.

Part II of the Instructor's Resource Manual contains sample quizzes with answers for every chapter. These quizzes, containing more than 800 test items, have been included as time-savers.

Part III includes two final examinations with a selection of questions from every chapter. Answers to these questions are also included.

Part IV pertains to Laboratory Experiments for Basic Chemistry, Fifth Edition. It provides detailed discussions of the experiments, including our own suggestions and short cuts. It also contains possible answers to the questions posed in the laboratory manual. Where applicable, instructions are given for preparing the various materials needed to perform the laboratory experiments. The fifth edition of the laboratory manual also contains CHEM-DECK, our copyrighted card game that has helped so many of our students learn how to write the formulas of chemical compounds.

Part V includes a complete listing of the transparencies available with Basic Concepts of Chemistry, Fifth Edition. Table/figure numbers and titles are provided.

Several teaching tools have been built into the textbook and the laboratory manual to make learning easier for students. Each chapter in both the textbook and the laboratory manual begins with a series of learning goals, which help students identify the topics that should be learned from a given study unit. The learning goals do not introduce time constraints or overly strict performance standards, and they are structured more loosely than conventional behavioral objectives. We believe that the learning goals will provide a starting point from which you can set specific goals to meet the needs of your classes. We hope that these suggested learning goals will give you maximum latitude in establishing goals tailored specifically for a given chapter.

By adopting our textbook, laboratory manual, instructor's guide, and solutions manual, you have selected a program whose approach to chemistry is thorough yet non-threatening. We believe the fifth edition of Basic Concepts of Chemistry, its accompanying Solutions Manual and Study Guide, and Laboratory Experiments for Basic Chemistry provide a solid foundation for the study of preparatory chemistry.

Over the years we have received many helpful suggestions from those who have used our books. We continue to solicit your input so that we can make future editions even more effective and useful.

Alan Sherman
Sharon J. Sherman
Leonard Russikoff

UTILIZING COOPERATIVE LEARNING IN THE CHEMISTRY CLASSROOM

In the fifth edition of Basic Concepts of Chemistry we have focused on the inclusion of a variety of teaching methods that will help students learn better. One of the techniques we have incorporated is cooperative learning, which is widely described in current educational literature. Cooperative learning is a set of well researched instructional strategies that enables the instructor to vary the presentation of the course content. Employment of these techniques provides a format for experiential learning, collaborative endeavors, and critical thinking in the classroom. Instead of presenting content in the traditional lecture style followed by whole class question and answer, cooperative learning techniques allow for students to discuss the material while interacting with each other. While such interaction can take place in the classroom, it can also take place out of class in small study groups. Throughout the text you will find numerous problems marked for cooperative problem solving. These problems are suitable for discussions which center around key concepts in chemistry.

In the college or university setting, chemistry is often taught in a large lecture session followed by smaller recitation sessions. Students often spend time in the recitation sessions reviewing material and asking questions. Cooperative learning can be used in the recitation sessions by breaking the class into teams of four or five students. The students work together on problems that reinforce the content presented in the lecture. Instead of having the instructor responsible for explanations and solutions, the students work together to discuss the content that was presented and solve the problems that are found both in the chapter and at the end of the chapter. The instructor circulates around the room answering questions when necessary, but students are responsible for their own learning and for the learning of the other students in the group. This technique fosters cooperation among students. It also allows students to share their thoughts regarding the content. When students verbalize incorrect information their ideas can be clarified by the group. The instructor is present to provide assistance when needed.

When the course takes place in a setting where the class size is smaller, there are additional ways to utilize cooperative learning. The instructor may choose to present material in lecture style, stopping where appropriate to have groups of three or four students confer and discuss the material. The group can generate questions about the content which need clarification. Again, students can work in small groups to solve problems. Discussion can take place in the groups, giving students the opportunity to verbalize their thoughts and clarify their thinking.

Study groups are another important asset to the chemistry student. Small groups of students should be encouraged to work together out of class. The problems marked for cooperative problem solving can be used in the study group. Any unanswered questions can be directed to the instructor at the next class. Besides providing an opportunity for discussion and reinforcement of skills and ideas, the cooperative group provides an important social experience for the beginning chemistry student. The camaraderie that develops in the group offers support for the student who is anxious about tackling chemistry.

UTILIZING COOPERATIVE LEARNING IN THE CHEMISTRY CLASSROOM

We have successfully used cooperative learning strategies for several years. We would be pleased to have you share your successful instructional strategies with us so that we may incorporate them into future editions of Basic Concepts of Chemistry. Send your comments to Professor Alan Sherman, Department of Chemistry, Middlesex County College, 155 Mill Road, P.O. Box 3050, Edison, New Jersey 08818-3050.

PART I

MATHEMATICS PRETEST/RETEST

PART I

MATHEMATICS PRETEST/RETEST

PART A ARITHMETIC

1. The sum of 26 and 39 is:
   (a) 65 (b) 55 (c) 45 (d) 13

2. Subtract 17 from 44. The answer is:
   (a) 37 (b) 7 (c) 27 (d) 61

3. The difference between 85 and 38 is:
   (a) 57 (b) 37 (c) 53 (d) 47

4. Multiply 15 by 8. The answer is:
   (a) 23 (b) 120 (c) 40 (d) 80

5. Multiply 1.5 by 8. The answer is:
   (a) 12 (b) 2.3 (c) 1.2 (d) 8

6. The sum of 3.75 and 29.44 is:
   (a) 22.19 (b) 23.19 (c) 33.19 (d) 3.319

7. The difference between 24.31 and 18.62 is:
   (a) 6.79 (b) 42.93 (c) 6.69 (d) 5.69

8. When $20 \div 5$, the answer is:
   (a) 100 (b) 0.25 (c) 4 (d) 5

9. When $5 \div 20$, the answer is:
   (a) 0.25 (b) 100 (c) 4 (d) 0.20

10. When $2.5 \div 0.5$, the answer is:
    (a) 0.5 (b) 5 (c) 50 (d) 0.20

11. When $0.45 \div 0.9$, the answer is:
    (a) 0.5 (b) 2 (c) 5 (d) 0.05

12. Multiply $\frac{2}{5}$ by $\frac{25}{40}$. The product is:
    (a) 0.50 (b) 0.125 (c) 0.25 (d) 0.40

13. When $\frac{2}{9} \div \frac{14}{45}$, the answer is:
(a) 0.71 (b) 1.8 (c) 0.56 (d) 0.36

14. Subtract $\frac{5}{8}$ from $\frac{3}{4}$. The answer is:
(a) 0.25 (b) 0.13 (c) 0.63 (d) 0.50

15. Add the following: $\frac{1}{9} + \frac{1}{3} + \frac{1}{6}$. The answer is:
(a) 0.27 (b) 0.17 (c) 0.61 (d) 0.33

16. When $6 \div \frac{3}{4}$, the answer is:
(a) 8 (b) 6 (c) 4.5 (d) 12

PART B PERCENTAGES

1. What percentage of 15 is 6?
(a) 10 (b) 20 (c) 30 (d) 40

2. Given that a class has 40 students and that 30% are men, the number of men in the class is:
(a) 28 (b) 12 (c) 14 (d) 16

3. A family has an income of $120 a week and spends $40 on food. The percentage spent on food is:
(a) 33 (b) 40 (c) 60 (d) 12

4. Suppose that you get 20 questions right on a test that contains 50 questions. The percentage right is:
(a) 20 (b) 30 (c) 40 (d) 60

5. Your gross pay is $100, but you take home only $80. The percentage deducted is:
(a) 10 (b) 20 (c) 30 (d) 40

PART C MEASURING IN MATH

1. The formula used to find the area of a circle is:
(a) πr^2 (b) πd (c) πd^2 (d) πr

2. The radius of a circle is r. The circumference is:
(a) πr^2 (b) πr (c) $2\pi r$ (d) r^2

3. When the sides of a rectangle are 3 ft and 5 ft, the area is:
(a) 8 sq ft (b) 1.7 sq ft (c) 7.5 sq ft (d) 15 sq ft

4. When the legs of a right triangle are 3 ft and 4 ft, the length of the hypotenuse is:
(a) 3 ft (b) 4 ft (c) 5 ft (d) 7 ft

5. The area of the right triangle of Question 4 is:
(a) 3 sq ft (b) 6 sq ft (c) 4 sq ft (d) 12 sq ft

6. When the edge of a cube is 4 in., its volume is:
(a) 64 cu in. (b) 12 cu in. (c) 16 cu in. (d) 32 cu in.

7. A tin can which has a radius of 2 in. and a height of 6 in. has a volume of:
(a) 24 cu in. (b) 75 cu in. (c) 72 cu in. (d) 36 cu in.

PART D EXPONENTS

1. The product of 10^{-2} and 10^{-4} is:
(a) 10^{-3} (b) 10^{-4} (c) 10^{-6} (d) 10^{6}

2. When $10^3 \div 10^2$, the answer is:
(a) 10 (b) 10^{-5} (c) 10^5 (d) 10^{-1}

3. In scientific notation, one writes 237,000 as:
(a) 23.7 (b) 2.37×10^5
(c) 2.37×10^{-5} (d) 2.37×10

4. Given that $n = \frac{(2 \times 10^4)(3 \times 10^{-6})}{1 \times 10^{-8}}$, find n.
(a) 5×10^{-6} (b) 6×10^2
(c) 6×10^{-6} (d) 6×10^6

5. Find $\sqrt{25x^2y^4}$.
(a) $5xy$ (b) $25xy^2$ (c) $5xy^2$ (d) $10xy^2$

6. What is $\sqrt{6.4 \times 10^5}$?
(a) 8×10^3 (b) 8×10^6
(c) 3.2×10^2 (d) 8×10^2

7. Given that $n = 4$, then n^2 is equal to:
(a) 16 (b) 12 (c) 8 (d) 20

PART E ALGEBRA

1. When you are told that $A = \frac{1}{2}(bh)$, you know that b is equal to:
(a) $\frac{h}{2A}$ (b) $2Ah$ (c) $\frac{A}{2h}$ (d) $\frac{2A}{h}$

2. Given $mv^2 = \frac{e^2}{r}$. When you solve for m, the result is:
(a) $\frac{e^2}{rv^2}$ (b) $\frac{rv^2}{e^2}$ (c) $\frac{e^2v^2}{r}$ (d) $\frac{v^2}{e^2r}$

3. Solve the following equation for x: $6x - 4 = 3x + 8$
   (a) 2 (b) 4 (c) 6 (d) 8

4. You are told that Y is inversely proportional to X, and that Y is 10 when X is 6. When Y is 20, X is:
   (a) 12 (b) 3 (c) 6 (d) 8

5. In the equation $D = \frac{m}{V}$, where D is density, m is mass and V is volume, what would V equal in terms of m and D?
   (a) mD (b) $\frac{m}{D}$ (c) $\frac{D}{m}$ (d) $m - D$

PART II

CHAPTER QUIZZES WITH ANSWERS

PART II

CHAPTER QUIZZES WITH ANSWERS

CHAPTER 1 Quiz

1. List the four elements that Empedocles thought made up the world.

2. What was the basic aim of chemistry from A.D. 300 to A.D. 1600?

3. The modern age of chemistry dawned with the publication of Robert Boyle's book ________________. (Name the book)

4. List two contributions of the Egyptians to the development of chemistry.

5. What important idea did Democritus establish about the nature of things?

6. While the Greeks were studying philosophy and mathematics, the Egyptians were already practicing the art of chemistry, which they called ________________.

7. The people who tried to turn lead into gold were known as ________________.

8. True or False: The Egyptians mined and purified the metals gold, silver, and antimony.

9. During the 1700's and early 1800's, most chemists believed that there were two main branches of chemistry. Name them.

10. Name the two classes of chemical compounds.

11. Which of the following is science?
    (a) a doctor performs a heart transplant
    (b) subatomic particles are discovered to be part of all atoms
    (c) a wristwatch TV is made available to the public
    (d) electric automobiles are produced

12. A simple statement or mathematical equation describing some basic fact or relationship of nature is called a(n)
    (a) experiment (b) hypothesis
    (c) law (d) theory

13. Which of the following is an example of technology?
    (a) subatomic particles are found to be part of all atoms
    (b) the law of conservation of mass is discovered
    (c) energy levels in atoms are discovered
    (d) a doctor performs a kidney transplant

14. A tentative explanation of a pattern or regularity is called a(n)
    (a) experiment (b) hypothesis
    (c) law (d) theory

15. Which of the following is science?
    (a) a doctor performs a kidney transplant
    (b) protons are found to be composed of even smaller subatomic particles
    (c) a light bulb that lasts ten years is produced
    (d) a nuclear fusion reactor is built

16. A detailed explanation that is useful in helping us describe and organize scientific knowledge is called a(n)
    (a) experiment
    (b) hypothesis
    (c) law
    (d) theory

17. Which of the following is an example of technology?
    (a) the law of conservation of energy is discovered
    (b) the nuclear model of the atom is proposed
    (c) solid rocket propellants are produced
    (d) the effect of the drug tPA on heart action is studied

18. An observation of some natural phenomenon that is carried out under controlled conditions so that it is possible to duplicate the results and draw rational conclusions is known as a(n)
    (a) experiment
    (b) hypothesis
    (c) law
    (d) theory

19. Finding knowledge for the sake of knowing more about our universe is called:
    (a) basic science
    (b) technology
    (c) applied science
    (d) earth science

20. Finding knowledge for the purpose of building a new type of computer is called:
    (a) basic science
    (b) technology
    (c) applied science
    (d) earth science

CHAPTER 1 Quiz Answers

1. Earth, air, fire, and water

2. The conversion of base metals into gold and the search to find the "elixir of life" (that is, alchemy)

3. The Sceptical Chymist

4. Mining and purifying metals, such as gold, silver, and copper, and making embalming fluids and dyes

5. The concept of the atom

6. khemia

7. alchemists

8. False

9. Organic chemistry and inorganic chemistry

10. Organic and inorganic

11. b	12. c	13. d	14. b	15. b
16. d	17. c	18. a	19. a	20. c

CHAPTER 2 Quiz

1. Fill in the blanks.

   ____________ kg = ____________ g = 250 dg = ____________ cg

   8 km = ____________ m = ____________ cm = ____________ mm

   25 mL = ____________ L = ____________ cm^3

2. (a) Write the following numbers in scientific notation:

   803,000 0.00725 2650

   (b) Perform the indicated operation and give the answer in scientific notation.

   (20,000)(3,000) = $\frac{0.006}{0.02} =$

3. Perform the following temperature conversions:

   (a) $1\bar{0}$°F = ________ °C (b) $2\bar{0}$°C = ________ °F

4. A block of metal weighs 60.0 g and has dimensions of 1.0 cm × 3.0 cm × 5.0 cm. Find the density of the block.

5. A substance has a density of 2.0 g/cm^3 and weighs 50.0 g. What is the volume of the substance?

6. Find the area of a rectangle that measures 11.0 m by 3.00 m. Report your answer to the proper number of significant figures.

7. Determine the volume of a metal cylinder that has a radius of 6.00 cm and a height of 20.0 cm. Report your answer to the proper number of significant figures. (Hint: $V = \pi r^2 h$)

8. Fill in the blanks.

   ____________ km = 18.40 m = ____________ dm = ____________ cm

   = ____________ mm

9. Convert 85.0 mL into liters.

10. There are 3.28 ft in 1.00 m. How many feet are there in 15.0 m?

11. Express $2\overline{000}$ lb in kilograms and in grams. (Hint: 454 g = 1.00 lb)

12. A cube measures 5.00 cm on each side and has a mass of $75\overline{0}$ g. What is the density of the cube?

13. Calculate the mass of a rectangular solid that measures 2.00 cm by 3.00 cm × 8.00 cm and has a density of 2.50 g/cm^3.

14. Calculate the volume of 432 g of ethyl alcohol. (Hint: The density of ethyl alcohol is 0.800 g/cm^3.)

15. (a) Write the following numbers in scientific notation:

    0.000081   9,500,000,000

    (b) Perform the indicated operation and give the number in scientific notation.

    $$(9{,}000{,}000)(0.0002) = \frac{4{,}000{,}000{,}000}{0.0008} =$$

16. Perform the following temperature conversions:
    (a) –94.0°F = ?°C   (b) 80.0°C = ?°F

17. Determine the length of a cube that has a density of 9.00 g/cm^3 and a mass of 576 g.

18. A solid metal object weighs $15\overline{0}$ g. The object is placed in a graduated cylinder containing water. The initial volume of the graduated cylinder was 10.0 cm^3, and the final water level of the graduated cylinder, after immersion of the object, is 40.0 cm^3. What is the density of the object?

19. Express 60.0 miles/h in feet per second and in meters per second. (Hint: 1 m = 3.28 ft)

20. Water has a density of 62.4 lb/ft^3. Express this in grams per cubic centimeter. Show your work. (Hint: 1.00 lb = 454 g, and 1.00 cm^3 = 0.0000352 ft^3)

21. Which of the following numbers indicates three significant figures?
    (a) 480   (b) 480.0
    (c) 4.80×10^2 (d) 4.8×10^2

22. The number 0.007003 has ____________ significant figures.
    (a) 1   (b) 2   (c) 3   (d) 4

23. Using significant figures, calculate the area of a square whose sides are 1.2 cm.
    (a) 1.44 cm^2 (b) 1.4 cm^2
    (c) 1.40 cm^2 (d) 5.6 cm^2

24. Using significant figures, find the perimeter of a rectangle whose sides are 9.46 cm and 15.2 cm.
    (a) 49.3 cm   (b) 49.32 cm
    (c) 24.6 cm   (d) 24.66 cm

25. The number 8,000,000 may be expressed in scientific notation as:
    (a) 8×10^5 (b) 8×10^6
    (c) 8×10^3 (d) 8×10^7

26. The number 0.00102 may be expressed in scientific notation as:
    (a) 1.02×10^{-3} (b) 102×10^{-4}
    (c) 1.02×10^{3} (d) 10.2×10^{-5}

27. The area of a table that measures 1.50 m by 2.00 m is:
    (a) 3.50 m^2 (b) 7.00 m^2
    (c) 3.0 m^2 (d) 3.00 m^2

28. The volume of a cube that measures 4.50 cm on each side is:
    (a) 20.25 cm^3 (b) 20.3 cm^3
    (c) 91.1 cm^3 (d) 91.125 cm^3

29. The volume of a cylinder that has a radius of 4.00 cm and a height of 10.0 cm is:
    (a) 502 cm^3 (b) 126 cm^3
    (c) 14.0 cm^3 (d) 40.0 cm^3

30. The volume of a sphere that has a radius of 5.00 cm is: (<u>Hint</u>: the formula for the volume of a sphere is $\frac{4}{3}\pi r^3$).
    (a) 6.67 cm^3 (b) 523 cm^3
    (c) 20.9 cm^3 (d) 532 cm^3

31. The number of grams in 0.160 kg is:
    (a) 16.0 g (b) 160 g
    (c) 1600 g (d) 1.6 g

32. The number of mL in 14.00 dl is:
    (a) 0.1400 mL (b) 140 mL
    (c) 1.4 mL (d) 1400 mL

33. If 1.00 inch = 2.54 cm, how many inches are in 30.5 cm?
    (a) 12.0 inches (b) 6.00 inches
    (c) 76.2 inches (d) 38.1 inches

34. If 1.00 pound = 454 g, how many pounds are in 22.7 g?
    (a) 10,300 pounds (b) 20.0 pounds
    (c) 2.00 pounds (d) 0.0500 pound

35. If 1.00 L = 1.06 quarts, how many liters are there in 4.00 quarts?
    (a) 3.77 L (b) 4.24 L
    (c) 0.265 L (d) 2.65 L

36. If 1.00 m = 3.28 feet, how many meters are there in 10.0 feet?
    (a) 32.8 m (b) 3.05 m
    (c) 3.28 m (d) 0.328 m

37. A block of aluminum has a mass of 10.0 g. The density of aluminum is 2.70 g/cc. What is the volume of the block?
    (a) $27\overline{0}$ cc (b) 0.270 cc
    (c) 12.7 cc (d) 3.70 cc

38. What is the mass of a cube that has a density of 8.20 g/cc, and volume of 125 cc?
    (a) 1030 g (b) 103 g
    (c) 15.2 g (d) 152 g

39. A temperature of 20.0°C is equivalent to:
(a) 40.0°F (b) 60.0°F
(c) 68.0°F (d) 93.6°F

40. A temperature of 54.0°F is equivalent to:
(a) 27.0°C (b) 129°C
(c) 12.2°C (d) 20.0°C

CHAPTER 2 Quiz Answers

1. 0.025 kg = 25 g = 250 dg = 2500 cg
8 km = 8000 m = 800,000 cm = 8,000,000 mm
25 mL = 0.025 L = 25 cm^3

2. (a) 8.03×10^5 7.25×10^{-3} 2.65×10^3

(b) 6×10^7 3×10^{-1}

3. (a) –12°C (b) 68°F

4. 4.0 g/cm^3

5. 25 cm^3

6. 33.0 m^2

7. 2260 cm^3

8. 0.01840 km = 18.40 m = 184.0 dm = $184\bar{0}$ cm = $18{,}4\bar{0}0$ mm

9. 0.0850 L

10. 49.2 ft

11. $908{,}\bar{0}00$ g = 908.0 kg

12. 6.00 g/cm^3

13. $12\bar{0}$ g

14. 540 cm^3

15. (a) 8.1×10^{-5} 9.5×10^9

(b) 1.8×10^3 5×10^{12}

16. (a) –70°C (b) 176°F

17. 4.00 cm

18. 5.00 g/cm^3

19. 88.0 ft/s, 26.8 m/s

20. 0.997 g/cm^3

21. c	22. d	23. b	24. a	25. b
26. a	27. d	28. c	29. a	30. c
31. b	32. d	33. a	34. d	35. a
36. b	37. d	38. a	39. b	40. c

CHAPTER 3 Quiz

1. Match each word on the left with its definition on the right.
   (a) Homogeneous
   (b) Heterogeneous
   (c) Mixture
   (d) Compound
   (e) Element

   (1) The basic building block of matter
   (2) The word used to describe matter that is uniform throughout
   (3) A type of matter in which each part retains its own properties
   (4) A chemical combination of two or more elements
   (5) The word used to describe matter that is not uniform throughout

2. Determine whether each of the following processes involves chemical or physical properties:
   (a) Ice melts.
   (b) Sugar dissolves in water.
   (c) Milk sours.
   (d) Eggs become rotten.
   (e) Water boils.
   (f) An egg is hard-cooked.

3. (a) What is the smallest particle of matter that can enter into a chemical combination?
   (b) What is the smallest uncharged individual unit of a compound that is composed of two or more different atoms?

4. Classify each of the following elements as metal, metalloid, or nonmetal:
   (a) Mn (b) Nd (c) Al (d) At
   (e) Pt (f) Cl (g) Ra

5. Explain the difference between Co and CO.

6. What is the molecular mass of a compound?

7. What is the formula mass of a compound?

8. If in the periodic table oxygen were assigned an atomic mass of 1, what would be the atomic mass of sulfur?

9. Name the elements present in each of the following compounds:
   (a) $MgCl_2$ (b) N_2O
   (c) $(NH_4)_2SO_4$ (d) H_3PO_4

10. Write the chemical formula of each of the following, given the number of atoms in a molecule or formula unit of the compound:
    (a) one nitrogen atom, two oxygen atoms (nitrogen dioxide)
    (b) two sodium atoms, one sulfur atom (sodium sulfide)
    (c) three potassium atoms, one arsenic atom, four oxygen atoms (potassium arsenate)
    (d) two phosphorus atoms, five oxygen atoms (diphosphorus pentoxide)

11. Classify each of the following as an element, compound, or mixture:
    (a) gold (b) air
    (c) carbon dioxide (d) wine
    (e) table salt

12. State whether each of the following involves a physical or chemical change:
    (a) toasting bread (b) water freezing
    (c) tearing paper (d) burning wood

13. Determine the molecular or formula mass of each of the following compounds. (For this exercise, round all atomic masses to one decimal place.)
    (a) OsO_4 (b) HNO_3
    (c) $Fe(OH)_2$ (d) $Ba_3(PO_4)_2$

14. Explain the difference between Hf and HF.

15. Determine the molecular or formula mass of each of the following compounds. (For this exercise, round all atomic masses to one decimal place.)
    (a) Al_2O_3 (b) $CuBr_2$
    (c) H_3PO_4 (d) LiCl
    (e) $Ca(OH)_2$ (f) $Fe_2(SO_4)_3$
    (g) $(NH_4)_2S$ (h) $CoCl_2$

16. Distinguish between heterogeneous and homogeneous.

17. State the Law of Definite Composition.

18. Describe the difference between an atom and a molecule.

19. Distinguish between an element and a compound.

20. Complete the sentence:
    "Matter can neither be created nor destroyed, it can change from one form to another in a normal chemical reaction" is a statement of the
    ____________________.

21. Which of the following substances is not an element?
    (a) mercury (b) gold
    (c) arsenic (d) carbon monoxide

22. What is the total number of atoms in one molecule of $C_{12}H_{22}N_2O_8S_2$?
    (a) 1 (b) 5 (c) 46 (d) 92

23. Based on the atomic mass scale that we use today, how many times heavier is an atom of copper compared to an atom of oxygen?
    (a) 1 (b) 2 (c) 3 (d) 4

24. Which of the following is <u>not</u> a compound?
(a) sucrose
(b) sodium
(c) water
(d) calcium chloride

25. What is the total number of atoms in one molecule of $C_{12}H_{22}O_{11}$?
(a) 1
(b) 6
(c) 45
(d) 18

26. In the periodic table, if bromine were assigned an atomic mass of five, what would be the atomic mass of oxygen?
(a) 5
(b) 3
(c) 2
(d) 1

27. Which of the following substances is <u>not</u> an element?
(a) carbon
(b) hydrogen
(c) water
(d) sulfur

28. What is the total number of atoms in one molecule of $C_{21}H_{30}Cl_2N_4O_3$?
(a) 1
(b) 10
(c) 60
(d) 180

29. Based on the atomic mass scale that we used today, how many times heavier is an atom of calcium compared to an atom of helium?
(a) 1
(b) 2
(c) 4
(d) 10

30. A pure substance composed of two or more elements chemically combined is a(n)
(a) atom
(b) compound
(c) mixture
(d) solution

31. Which of the following substances is not homogeneous?
(a) water
(b) iron
(c) sugar
(d) sand

32. Iron and sulfur are ground together using a mortar and pestle. What results is a(n):
(a) mixture
(b) compound
(c) element
(d) molecule

33. Matter that has different parts with different properties is said to be:
(a) homogeneous
(b) composed of a pure substance
(c) heterogeneous
(d) a compound

34. Table salt is an example of a(n):
(a) element
(b) homogeneous mixture
(c) compound
(d) heterogeneous mixture

35. The smallest part of an element that can enter into chemical reactions is the
(a) molecule
(b) atom
(c) compound
(d) solution

36. Ethyl alcohol combined with water is an example of a(n):
(a) element
(b) homogeneous mixture
(c) compound
(d) heterogeneous mixture

37. Mercury(II) oxide is an example of a(n):
(a) molecule
(b) atom
(c) compound
(d) solution

38. Helium is an example of a(n):
    (a) element
    (b) homogeneous mixture
    (c) compound
    (d) heterogeneous mixture

39. Air is an example of a(n):
    (a) molecule
    (b) atom
    (c) compound
    (d) solution

40. Which of the following is an element?
    (a) CO (b) Co (c) SO_2 (d) H_2O

41. If oxygen were assigned an atomic weight of one, what would be the atomic weight of copper relative to oxygen?
    (a) 2 (b) 3 (c) 4 (d) 5

42. Which of the following is an element?
    (a) NI (b) Ni (c) CO_2 (d) Cl_2O

43. If silicon were assigned an atomic weight of one, what would be the atomic weight of calcium relative to silicon?
    (a) 2 (b) 3 (c) 4 (d) 5

CHAPTER 3 Quiz Answers

1. (a) 2 (b) 5 (c) 3 (d) 4 (e) 1

2. (a) physical
   (b) physical
   (c) chemical
   (d) chemical
   (e) physical
   (f) chemical

3. (a) atom
   (b) molecule

4. (a) metal
   (b) metal
   (c) metal
   (d) metalloid
   (e) metal
   (f) nonmetal
   (g) metal

5. Co is the symbol for the element cobalt. CO is the chemical formula for carbon monoxide.

6. The molecular mass of a compound is the sum of the atomic masses that compose a molecule of the compound.

7. The formula mass of a compound is the sum of the atomic masses that compose a formula unit of the compound.

8. The atomic mass of sulfur would be 2.

9. (a) magnesium and chlorine
   (b) nitrogen and oxygen
   (c) nitrogen, hydrogen, sulfur, and oxygen
   (d) hydrogen, phosphorus, and oxygen

10. (a) NO_2 (b) Na_2S (c) K_3AsO_4 (d) P_2O_5

11. (a) element (b) mixture
    (c) compound (d) mixture
    (e) compound

12. (a) chemical (b) physical
    (c) physical (d) chemical

13. (a) 254.2 (b) 63.0 (c) 89.8 (d) 601.9

14. Hf is the symbol for the element hafnium, and HF is the formula for hydrogen fluoride.

15. (a) 102.0 (b) 223.3 (c) 98.0 (d) 42.4 (e) 74.1
    (f) 399.9 (g) 68.1 (h) 129.9

16. Homogeneous matter is matter that has the same composition throughout. Heterogeneous matter is matter that has different parts with different properties.

17. The Law of Definite Composition states that a given compound always has the same elements in the same proportions by mass.

18. An atom is the smallest part of an element that retains the physical and chemical properties of that element. A molecule is the smallest part of a compound that retains the physical and chemical properties of that compound. Molecules are made up of two or more atoms bonded covalently. They may be identical.

19. Elements are the basic building blocks of matter. Compounds are chemical combinations of two or more elements.

20. Law of Conservation of Mass

21. d 22. c 23. d 24. b 25. c

26. d 27. c 28. c 29. d 30. b

31. d 32. a 33. c 34. c 35. c

36. b 37. c 38. a 39. d 40. b

41. c 42. b 43. c

CHAPTER 4 Quiz

1. Determine the number of protons, electrons, and neutrons in:
   (a) $^{28}_{13}Al$ (b) $^{14}_{6}C$ (c) $^{1}_{1}H$

2. Complete the following table.

Particle	Charge	Atomic mass (amu, approx.)
Proton		
		0
	0	

3. How many atoms of ^{28}Si are present in 1.00×10^6 silicon atoms, given that the percentage of ^{28}Si is 92.2 percent?

4. What is the relationship between the atomic number and the number of electrons in a neutral atom?

5. Consider the following unknown atoms:

   $^{200}_{80}A$ $\quad$ $^{208}_{82}B$ $\quad$ $^{222}_{86}C$ $\quad$ $^{184}_{74}D$

   (a) Which neutral atom has the most electrons?
   (b) Which element has the least neutrons?

6. Match each scientist with his model of the atom:
   (a) Thomson $\quad$ (1) Nuclear atom model
   (b) Rutherford $\quad$ (2) Plum-pudding model

7. Determine the number of protons, electrons, and neutrons in neutral atoms of:
   (a) $^{32}_{16}S$ $\quad$ (b) $^{31}_{15}P$ $\quad$ (c) $^{79}_{35}Br$ $\quad$ (d) $^{81}_{35}Br$

8. If you could turn lead into platinum, how many protons, electrons, and neutrons would you remove from a neutral $^{208}_{82}Pb$ atom to turn it into a neutral $^{195}_{78}Pt$ atom?

9. You are given the following information for three unknown atoms:

   Atom X has 25p, 25e, and 30n.
   Atom Y has 26p, 26e, and 30n.
   Atom Z has 25p, 25e, and 31n.

   Which atom (X, Y, or Z) is not an isotope of the other two?

10. Complete the following table.

Symbol	Protons	Electrons	Neutrons	Mass No.	Atomic No.
$^{13}_{6}C$					
	10			20	
		15	16		
$_{92}U$			143		

11. Write the standard isotopic notation for each of the following elements:
    (a) 55p, 55e, 68n $\quad$ (b) 14p, 14e, 16n

12. Complete the sentence.

    Atoms of an element that have the same number of electrons and protons, but different numbers of neutrons, are called ______________.

13. Match each term on the left with its definition on the right.
    (a) Atomic number $\quad$ (1) The number of protons or electrons in a neutral atom
    (b) Mass number $\quad$ (2) The sum of the protons and neutrons in an atom

14. Element Z exists in three isotopic forms with the following abundances:

    50.0 percent ^{100}Z, 40.0 percent ^{102}Z, and 10.0 percent ^{105}Z.

    Calculate the atomic mass of element Z.

15. If you have 5.000×10^9 Mg atoms, how many are ^{24}Mg atoms? (Hint: The percent abundance of ^{24}Mg is 78.70 percent.)

16. How many carbon atoms would one need to collect in order to obtain 2.00×10^6 ^{13}C atoms? (Hint: The percent abundance of ^{13}C is 1.11 percent.)

17. Boron exists in two isotopic forms: ^{11}B and ^{10}B. The atomic mass of boron is 10.80. Calculate the percent abundance of each isotope. For this problem, use the mass numbers of each isotope as the exact masses.

18. Consider the following unknown atoms:

    $^{300}_{110}A$ $^{310}_{112}B$ $^{315}_{114}C$

    (a) Which atom has the most protons?
    (b) Which atom has the most neutrons?
    (c) What is the mass number of element B?
    (d) Which atom has the fewest protons?

19. The element silicon exists as three isotopes with the following percent abundances:

    ^{28}Si = 92.21 percent ^{29}Si = 4.70 percent ^{30}Si = 3.09 percent

    Calculate the atomic mass of Si.

20. How many ^{25}Mg atoms are in 6.02×10^{22} Mg atoms obtained at random? (Hint: The percent abundance of ^{25}Mg is 10.13 percent.)

21. Which of the following statements is true about isotopes of the same element?
    (a) they always have the same number of neutrons
    (b) they always have the same number of protons
    (c) they always have different numbers of electrons
    (d) they always have the same atomic mass

22. The isotope $^{238}_{92}U$ has:
    (a) 238 p (b) 238 n (c) 92 n (d) 146 n

23. An isotope that has 35 p, 35 e, and 46 n is:
    (a) $^{81}_{46}Pd$ (b) $^{81}_{35}Br$ (c) $^{46}_{35}Br$ (d) $^{35}_{46}Pd$

24. Isotopes of the same element always have:
    (a) the same number of neutrons
    (b) the same atomic mass
    (c) different numbers of protons
    (d) the same atomic number

25. Which of the following statements is not true about isotopes of the same element?
    (a) they always have the different numbers of neutrons
    (b) they always have the different numbers of protons
    (c) they always have the same atomic number
    (d) they always have different atomic masses

26. The isotope ${}^{131}_{53}I$ has:
(a) 131p (b) 131n (c) 78p (d) 78n

27. The atom ${}^{14}_{6}C$ has:
(a) 6p, 6e, 14n
(b) 14p, 14e, 6n
(c) 8p, 8e, 6n
(d) 6p, 6e, 8n

28. The nuclear model of the atom was proposed by:
(a) Dalton
(b) Thomson
(c) Rutherford
(d) Bohr

29. The subatomic particle that has a negative one charge is called the:
(a) proton
(b) electron
(c) neutron
(d) neutrino

30. The isotope ${}^{106}_{46}Pd$ has:
(a) 46n (b) 106n (c) 60p (d) 60n

31. The "indivisible" model of the atom was proposed by:
(a) Dalton
(b) Thomson
(c) Rutherford
(d) Bohr

32. The subatomic particle that has a positive one charge is called the:
(a) proton
(b) electron
(c) neutron
(d) neutrino

33. The isotope ${}^{206}_{82}Pb$ has:
(a) 206n (b) 124n (c) 82n (d) 206p

34. The "plum-pudding" model of the atom was proposed by:
(a) Thomson
(b) Dalton
(c) Rutherford
(d) Bohr

35. The subatomic particles that have similar masses are:
(a) protons and electrons
(b) protons and neutrons
(c) electrons and neutrons
(d) protons, electrons, and neutrons

36. The isotope ${}^{21}_{10}Ne$ has:
(a) 21n (b) 21p (c) 11n (d) 10n

37. The mass number of an atom is the sum of the masses of the:
(a) protons + neutrons
(b) protons + electrons
(c) protons
(d) electrons

38. The isotope ${}^{222}_{86}Rn$ has:
(a) 222p (b) 222n (c) 86p (d) 86n

39. The Rutherford model of the atom is also known as the:
(a) plum-pudding model
(b) nuclear model
(c) energy level model
(d) quantum model

40. The number of neutrons in an atom is:
    (a) the same as the atomic number
    (b) the same as the mass number
    (c) the difference between the mass number and atomic number
    (d) the sum of the mass number and atomic number

41. The isotope $^{251}_{98}Cf$ has:
    (a) 98n (b) 251p (c) 251n (d) 153n

42. The number of electrons in a neutral atom is:
    (a) the same as the mass number
    (b) the same as the atomic number
    (c) the difference between the mass number and atomic number
    (d) the sum of the mass number and atomic number

43. The isotope $^{255}_{102}No$ has:
    (a) 102n (b) 102p (c) 255n (d) 153p

CHAPTER 4 Quiz Answers

1. (a) 13p, 13e, 15n (b) 6p, 6e, 8n (c) 1p, 1e, 0n

2.

Proton	+1	1
Electron	–1	0
Neutron	0	1

3. 9.22×10^5 atoms

4. They are the same.

5. (a) C (b) D

6. (1) b, (2) a

7. (a) 16p, 16e, 16n (b) 15p, 15e, 16n
   (c) 35p, 35e, 44n (d) 35p, 35e, 46n

8. Remove 4p, 4e, and 9n.

9. Y is not an isotope of X and Z.

10.

Symbol	Protons	Electrons	Neutrons	Mass No.	Atomic No.
$^{13}_{6}C$	6	6	7	13	6
$^{20}_{10}Ne$	10	10	10	20	10
$^{31}_{15}P$	15	15	16	31	15
$^{235}_{92}U$	92	92	143	235	92

11. (a) $^{123}_{55}Cs$ (b) $^{30}_{14}Si$

12. isotopes

13. (1) a (2) b

14. 101.3

15. 3.935×10^{9} ^{24}Mg atoms

16. 1.80×10^{8} C atoms

17. ^{10}B = 19.6 percent, ^{11}B = 80.4 percent

18. (a) C (b) C (c) 310 (d) A

19. 28.1

20. 6.10×10^{21} ^{25}Mg atoms

21. b 22. d 23. b 24. d 25. b

26. d 27. d 28. c 29. b 30. d

31. a 32. a 33. b 34. a 35. b

36. c 37. a 38. c 39. b 40. c

41. d 42. b 43. b

CHAPTER 5 Quiz

1. Match the person on the left with that person's description of the atom on the right.
   (a) Dalton (1) The plum-pudding model
   (b) Rutherford (2) The indivisible atom
   (c) Thomson (3) The energy-level atom
   (d) Bohr (4) The nuclear atom

2. What subatomic particles are responsible for the line spectra of elements?

3. State the number of electrons in each energy level for the following atoms:
   (a) $_{8}O$ (b) $_{3}Li$ (c) $_{6}C$ (d) $_{10}Ne$

4. The Group IA elements have how many electrons in their outermost energy level?

5. What is the major difference between a continuous spectrum and a line spectrum?

6. Complete the sentence with the word "higher" or "lower."

   The farther away an energy level is from the nucleus, the ________________ the energy of an electron in that level.

7. What is the maximum number of electrons allowed in energy level 4, in level 5, and in level 6?

8. Match each word on the left with its definition on the right.

(a) Continuous spectrum	(1) Elements produce this
(b) Line spectrum	(2) Analyzes light
(c) Spectroscope	(3) The sun produces this

9. Fill in the words "ground state" or "excited state."

   A hydrogen atom has an electron in the M energy level. This electron is in an ________________ . A hydrogen atom has an electron in the K energy level. This electron is in the ________________ .

10. Complete the sentence with the word "spiral" or "jump."

    According to the Bohr model of the atom, electrons ________________ from one energy level to another.

11. Without using the periodic table, determine the electron configuration of each of the following elements:
    (a) $_{11}Na$ (b) $_{5}B$ (c) $_{17}Cl$

12. Determine how many electrons are in the outermost energy level of a neutral atom of:
    (a) Li (b) C (c) O

13. Using the periodic table, name the elements that have the following neutral electron configurations:
    (a) K has 2 electrons and L has 2 electrons.
    (b) K has 2 electrons, L has 8 electrons, and M has 1 electron.
    (c) K has 2 electrons, L has 8 electrons, and M has 5 electrons.

14. Complete the sentence.

    The number of electrons in the ________________ level determines the chemical properties of an element.

15. Write electron configurations for all the elements between 1 and 20 whose behavior is explained by the octet rule (in other words, those that have a stable octet of electrons).

16. Write electron configurations for all elements between 1 and 20 that are isoelectronic with sodium. (Hint: In this instance, "isoelectronic with" can be taken to mean "have the same number of electrons in the outermost energy level as.")

17. What can happen to an electron in an atom if it absorbs too much energy?

18. Show by writing electron configurations why the following elements should be in the indicated groups.
    (a) S (Group VIA) (b) F (Group VIIA)

19. What is chemically unique about the Group VIIIA elements?

20. How many electrons would it take to fill the K, L, M, and N energy levels?

21. The maximum number of electrons permitted in the L (or second) energy level is:
    (a) 2 (b) 6 (c) 8 (d) 18

22. The number of electrons in the outermost energy level of bromine is:
    (a) 7 (b) 6 (c) 5 (d) 4

23. The maximum number of electrons permitted in the M (or third) energy level is:
    (a) 2 (b) 6 (c) 8 (d) 18

24. The number of electrons in the outermost energy level of iodine is:
    (a) 7 (b) 6 (c) 5 (d) 4

25. The electron configuration for a chlorine atom ($_{17}Cl$) is:
    (a) K has 8, L has 8, M has 1
    (b) K has 2, L has 15
    (c) K has 2, L has 8, M has 7
    (d) K has 2, L has 8, M has 5, N has 2

26. The "energy level" model of the atom was proposed by:
    (a) Dalton (b) Thomson
    (c) Rutherford (d) Bohr

27. The maximum number of electrons permitted in the N (or fourth) energy level is:
    (a) 8 (b) 18 (c) 32 (d) 50

28. The maximum number of electrons permitted in the O (or fifth) energy level is:
    (a) 8 (b) 18 (c) 32 (d) 50

29. The maximum number of electrons permitted in the P (or sixth) energy level is:
    (a) 8 (b) 18 (c) 32 (d) 72

30. The model of the atom that was a refinement of the nuclear atom was proposed by:
    (a) Thomson (b) Dalton
    (c) Rutherford (d) Bohr

31. The atomic number of an atom that has a filled K and L energy level is:
    (a) 8 (b) 10 (c) 18 (d) 32

32. If the maximum number of electrons allowed in energy levels followed a $3n^2$ rule instead of a $2n^2$ rule, what would be the atomic number of the first noble gas?
    (a) 2 (b) 3 (c) 8 (d) 27

33. The Bohr model of the atom is also known as the:
    (a) plum-pudding model (b) nuclear model
    (c) energy level model (d) quantum model

34. If the maximum number of electrons allowed in energy levels followed a $3n^2$ rule instead of a $2n^2$ rule, what would be the atomic number of the second noble gas?
(a) 3 (b) 12 (c) 27 (d) 15

35. Which of the following energy sublevels does not exist?
(a) 1p (b) 2s (c) 3d (d) 4s

36. Which of the following energy sublevels does not exist?
(a) 3s (b) 3p (c) 3d (d) 3f

37. The atomic number of an atom that has a filled first and second energy level is:
(a) 2 (b) 8 (c) 10 (d) 18

38. The atoms of elements in Group VA of the periodic table all have in the outermost energy level:
(a) p^3 electrons (b) p^4 electrons
(c) p^5 electrons (d) p^6 electrons

39. The correct electron configuration for an atom of an element that has the atomic number of 20 is:
(a) $1s^22s^22p^63s^23p^63d^2$
(b) $1s^22s^22p^63s^23p^43d^4$
(c) $1s^22s^22p^63s^23p^64s^2$
(d) $1s^22s^22p^63s^23p^8$

40. An element whose atoms have the same number of electrons in their outermost energy level as atomic number 13 is:
(a) Ga (b) Ge (c) As (d) Se

CHAPTER 5 Quiz Answers

1. (1) c, (2) a, (3) d, (4) b

2. Electrons

3. (a) K has 2 and L has 6. (b) K has 2 and L has 1.
(c) K has 2 and L has 4. (d) K has 2 and L has 8.

4. One electron

5. In a continuous spectrum, one color merges into the next. In a line spectrum, dark bands separate the colors of light.

6. higher

7.

Energy level	Maximum number of electrons allowed
4	32
5	50
6	72

8. (1) b, (2) c, (3) a

9. excited state, ground state

10. jump

11. $_{11}Na$ — K has 2e, L has 8e, and M has 1e.
    $_{5}B$ — K has 2e and L has 3e.
    $_{17}Cl$ — K has 2e, L has 8e, and M has 7e.

12. (a) 1e (b) 4e (c) 6e

13. (a) Be (b) Na (c) P

14. outermost energy level

15. $_{10}Ne$ — K has 2e and L has 8e.
    $_{18}Ar$ — K has 2e, L has 8e, and M has 8e.

16. $_{1}H$ — K has 1e.
    $_{3}Li$ — K has 1e and L has 2e.
    $_{19}K$ — K has 2e, L has 8e, M has 8e, and N has 1e.

17. It can escape from the atom.

18. $_{16}S$ — K has 2e, L has 8e, and M has 6e.
    $_{9}F$ — K has 2e and L has 7e.

19. The Group VIIIA elements are unusually stable and tend not to react with the atoms of other elements.

20. 60

21. c	22. a	23. d	24. a	25. c
26. d	27. d	28. d	29. d	30. d
31. b	32. b	33. c	34. d	35. a
36. d	37. c	38. a	39. c	40. a

CHAPTER 6 Quiz

1. (a) In the periodic table, a vertical column is called a ______________.

   (b) In the periodic table, a horizontal row is called a ______________.

2. Which element has the largest atomic radius: P, Na, or Cl?

3. Which element has the smallest atomic radius: S, Se, or O?

4. Using subshell notation, write the electron configuration of $_{18}Ar$.

5. A certain neutral element has 2 electrons in the K shell, 8 electrons in the L shell, 18 electrons in the M shell, and 5 electrons in the N shell. List the following information for this element:
   (a) Atomic number
   (b) Total number of s electrons
   (c) Total number of p electrons
   (d) Total number of d electrons

6. Match each scientist with his contribution to chemistry.

(a) Döbereiner	(1) The law of octaves
(b) Newlands	(2) Modern periodic table
(c) Mendeleev	(3) Triads

7. Complete the sentence, placing the letters A and B in their proper places:

   In the periodic table the representative elements are known as the ______________ Group elements, and the transition metals are known as the ______________ Group elements.

8. List the following elements in order of increasing atomic radius:

   Cl, Ca, Sr, Rb, and S.

9. List the following elements in order of increasing ionization potential:

   Ba, Rb, Sr, and Mg.

10. Choose the correct answer: An S^{2-} ion is an S atom that has __________.
    (a) gained 2 protons   (b) lost 2 protons
    (c) gained 2 electrons   (d) lost 2 electrons

11. In what group of the periodic table do you find the most unreactive elements?

12. Choose the correct answer: An Al^{3+} ion is an aluminum atom that has

    ______________.
    (a) gained 3 protons   (b) lost 3 protons
    (c) gained 3 electrons   (d) lost 3 electrons

13. Electron affinity ______________ (increases/decreases) from left to right across a period of elements.

14. Using subshell notation, write the electron configuration of $_{21}Sc$.

15. According to the Bohr model of the atom, the f sublevel can hold 14 electrons. How many f orbitals are there in the 4th shell?

16. A certain neutral element has 2 electrons in the K level, 8 electrons in the L level, 18 electrons in the M level, and 7 electrons in the N level. List the following information for this element:
    (a) Atomic number
    (b) Total number of s electrons
    (c) Total number of p electrons
    (d) Total number of d electrons
    (e) Total number of f electrons
    (f) Number of protons

17. Use the $2n^2$ rule to calculate the maximum number of electrons allowed in the seventh energy level.

18. Which member of each pair has the larger radius?
    (a) Na or Na^{1+}
    (b) Cl or Cl^{1-}

19. Draw the shape of:
    (a) An s orbital
    (b) A p_x orbital
    (c) A p_y orbital
    (d) A p_z orbital

20. Write the electron configuration for the next noble gas following radon.

21. The scientist who first proposed triads of elements was:
    (a) Mendeleev
    (b) Döbereiner
    (c) Newlands
    (d) Dalton

22. The scientist who first devised the Law of Octaves was:
    (a) Mendeleev
    (b) Döbereiner
    (c) Newlands
    (d) Dalton

23. The scientist who first proposed and published a periodic table of elements was:
    (a) Mendeleev
    (b) Döbereiner
    (c) Newlands
    (d) Dalton

24. The elements sodium, magnesium, and aluminum are said to be in the same chemical:
    (a) family
    (b) group
    (c) period
    (d) triad

25. Which of the following elements have similar chemical properties?
    (a) Fe, Co, and Ni
    (b) Na, K, and Rb
    (c) As, Se, and Br
    (d) P, S, and Cl

26. The A-group elements are also known as:
    (a) representative elements
    (b) transition elements
    (c) inner transition elements
    (d) actinide elements

27. The B-group elements are also known as:
    (a) representative elements
    (b) transition elements
    (c) lanthanide elements
    (d) actinide elements

28. From among the following elements, the element with the largest atomic radius is:
    (a) Na (b) K (c) Rb (d) Cs

29. From among the following elements, the element with the lowest ionization potential is:
    (a) Mg (b) Ca (c) Sr (d) Ba

30. In the periodic table, elements with the lowest ionization potential are found in:
 (a) the lower left-hand corner
 (b) the lower right-hand corner
 (c) the upper left-hand corner
 (d) the upper right-hand corner

31. In the periodic table, the most unreactive elements are found in Group:
 (a) IA (b) IIIA (c) VIIA (d) VIIIA

32. A Br^{1-} ion is a Br atom that has:
 (a) gained a proton (b) lost a proton
 (c) gained an electron (d) lost an electron

33. A Na^{1+} ion is a Na atom that has:
 (a) gained a proton (b) lost a proton
 (c) gained an electron (d) lost an electron

34. Atoms of elements in Group VIA are most likely to form ions with a charge of:
 (a) 2– (b) 2+ (c) 6– (d) 6+

35. Atoms of elements in Group IIA are most likely to form ions with a charge of:
 (a) 2– (b) 2+ (c) 6– (d) 6+

36. The energy released when an additional electron is added to a neutral atom is known as the:
 (a) activation energy (b) electron affinity
 (c) enthalpy (d) ionization potential

37. The energy required to remove an electron from a neutral atom is known as the:
 (a) activation energy (b) electron affinity
 (c) enthalpy (d) ionization potential

38. An element with atomic number 114 should have chemical properties similar to:
 (a) gold (b) mercury
 (c) tellurium (d) lead

39. Among the following elements, the element with the highest ionization potential is:
 (a) Mg (b) S (c) Na (d) Cl

40. The total number of s electrons in an atom of the element potassium is:
 (a) 5 (b) 6 (c) 7 (d) 8

CHAPTER 6 Quiz Answers

1. (a) group or family (b) period

2. Na

3. O

4. $1s^2 2s^2 2p^6 3s^2 3p^6$

5. (a) Atomic number 33 (b) 8 (c) 15 (d) 10

6. (1) b, (2) c, (3) a

7. A, B

8. Cl, S, Ca, Sr, Rb
   smallest largest

9. Ba, Rb, Sr, Mg
   lowest highest

10. (c)

11. Group VIIIA, the noble gases

12. (d)

13. increases

14. $1s^22s^22p^63s^23p^64s^23d^1$

15. 7f orbitals

16. (a) atomic number = 35 (b) s electrons = 8 (c) p electrons = 17
    (d) d electrons = 10 (e) f electrons = 0 (f) protons = 35

17. 98 electrons

18. (a) Na (b) Cl^{1-}

19.

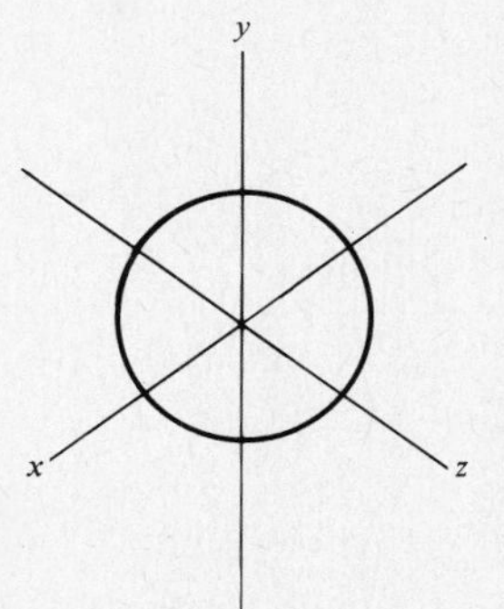

(a) s orbital

(b) p_x orbital

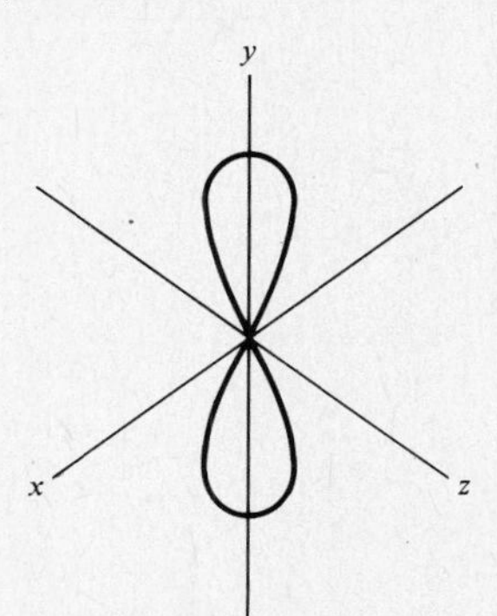

(c) p_y orbital

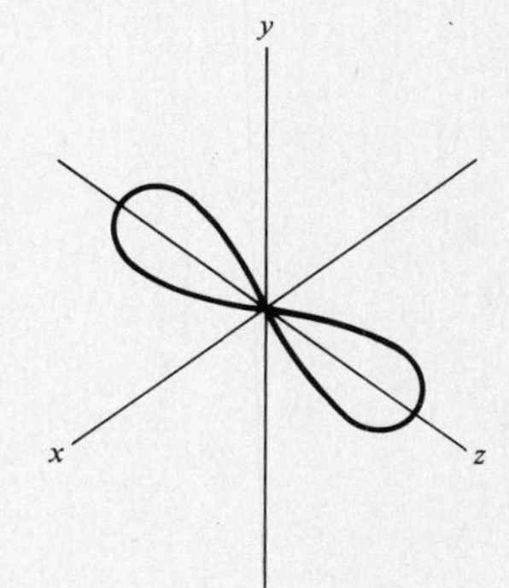

(d) p_z orbital

20. Element 118 would be the next inert gas after radon.

$1s^2 2s^2 2p^6 3s^2 3p^6 4s^2 3d^{10} 4p^6 5s^2 4d^{10} 5p^6 6s^2 4f^{14} 5d^{10} 6p^6 7s^2 5f^{14} 6d^{10} 7p^6$

21. b   22. c   23. a   24. c   25. b

26. a   27. b   28. d   29. d   30. a

31. d   32. c   33. d   34. a   35. b

36. b   37. d   38. d   39. d   40. c

CHAPTER 7 Quiz

1. Write the electron-dot structure for the following elements:
(a) Cs   (b) Mg   (c) S   (d) Al

2. Determine whether the bonds in the following compounds are polar or nonpolar.
(a) H_2O (b) CCl_4 (c) SF_6 (d) H_2

3. Determine whether the following molecules are polar or nonpolar.
(a) O=C=O   (b)

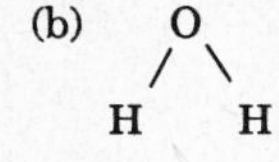

(c)

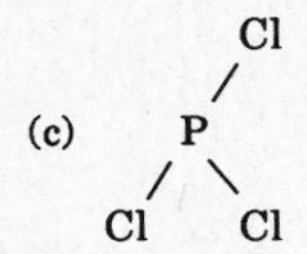

(d)

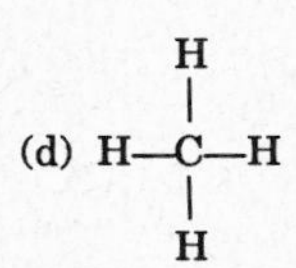

4. Determine the oxidation number of the underlined element or ion.
(a) $K\underline{Mn}O_4$ (b) $\underline{Pb}O_2$ (c) $Na\underline{Cl}O_4$ (d) $H\underline{N}O_3$

5. Determine the order of increasing bond polarity among the following groups:
O—O, N—O, C—O

6. How is an ionic bond different from a covalent bond?

7. Name the diatonic elements from memory.

8. When each atom supplies ______________ electrons to be shared, the bond formed is a double covalent bond.

9. Which of the diatomic elements contain:
(a) Single covalent bonds   (b) Double covalent bonds
(c) Triple covalent bonds

10. Write the electron-dot structure for each of the following compounds:
(a) HCN   (b) C_2H_4

11. Not all compounds follow the octet rule. With this in mind, write the electron-dot structure for each of the following compounds:
(a) PCl_5 (b) BF_3

12. Draw the bonding diagram for bromic acid, $HBrO_3$, which has three coordinate covalent bonds. (Hint: The bromine is bonded to the three oxygen atoms, and the hydrogen is bonded to one of the oxygen atoms.)

13. Determine whether each of the following compounds is bonded ionically or covalently.
(a) LiCl (b) NH_3 (c) MgF_2 (d) OF_2

(Hint: Electronegativity for Li is 1.0, for Cl 3.0, for N 3.0, for H 2.1, for Mg 1.2, for F 4.0, and for O 3.5.)

14. List the following compounds in terms of decreasing covalence:

H_2Se, H_2S, H_2O, H_2Te.

(Hint: Electronegativity for H is 2.1, for S 2.5, for Se 2.4, for O 3.5, and for Te 2.1.)

15. Draw the bonding diagram for the compound chloric acid, $HClO_3$, which has a coordinate covalent bond. (Hint: The chlorine is bonded to three oxygen atoms, and the hydrogen is bonded to one of the oxygen atoms.)

16. Which of the following compounds is the most ionic? Hf HCl HBr HI

17. Predict the compound that is most likely to be formed from each of the following pairs of elements:
(a) Mg and O (b) Al and O
(c) Cs and S

18. Find the oxidation number of the underlined element.
(a) $H_3\underline{P}O_4$ (b) $\underline{V}_2O_5$ (c) $\underline{Co}S$ (d) $\underline{U}F_6$

19. Determine whether each of the bonds in the following molecules is polar or nonpolar. Then determine whether the molecule itself is polar or nonpolar.

```
        Cl
        |
(a) Cl—C—Cl                    (b) S=C=S
        |
        Cl

(c)    N                       (d)    S
      /|\                            / \
     H H H                          H   H
```

(Hint: The electronegativity for C is 2.5, for Cl 3.0, for S 2.5, for N 3.0, and for H 2.1.)

20. Which of the following bonds is the most covalent?

C—H C—O N—O C—S

(Hint: The electronegativity for C is 2.5, for H 2.1, for O 3.5, for N 3.0, and for S 2.5.)

21. The carbon-carbon bond in a molecule of acetylene, C_2H_2, is a:
(a) single bond
(b) double bond
(c) triple bond
(d) ionic bond

22. Which of the following compounds has the most polar bonds?
(a) H_2O
(b) H_2S
(c) H_2Se
(d) H_2Te

23. The carbon-carbon bond in a molecule of ethylene, C_2H_4, is a(n):
(a) single bond
(b) double bond
(c) triple bond
(d) ionic bond

24. Which of the following compounds is the least polar?
(a) HF (b) HCl (c) HBr (d) HI

25. The carbon-oxygen bonds in a molecule of carbon dioxide, CO_2, are:
(a) single bonds
(b) double bonds
(c) triple bonds
(d) ionic bonds

26. Which of the following compounds has the most polar bonds?
(a) NH_3 (b) PH_3 (c) AsH_3 (d) SbH_3

27. Which of the following compounds is the most polar?
(a) HF (b) HCl (c) HBr (d) HI

28. The carbon-nitrogen bond in a molecule of hydrogen cyanide, HCN, is a(n):
(a) single bond
(b) double bond
(c) triple bond
(d) ionic bond

29. Which of the following compounds has a covalent bond?
(a) KCl (b) HCl (c) NaCl (d) RbCl

30. The carbon-sulfur bonds in a molecule of carbon disulfide (CS_2) are:
(a) single bonds
(b) double bonds
(c) triple bonds
(d) ionic bonds

31. Which of the following compounds has an ionic bond?
(a) SO_3 (b) HCl (c) NaCl (d) CO_2

32. The oxygen-fluorine bonds in a molecule of oxygen difluoride (OF_2) are:
(a) single bonds
(b) double bonds
(c) triple bonds
(d) ionic bonds

33. Which of the following compounds has an ionic bond?
(a) $CaCl_2$ (b) HCl (c) Cl_2 (d) NO_2

34. The nitrogen-nitrogen bond in a molecule of nitrogen gas (N_2) is a(n):
(a) single bond
(b) double bond
(c) triple bond
(d) ionic bond

35. Which of the following compounds has a nonpolar covalent bond?
(a) LiCl (b) OF_2 (c) NO_2 (d) Br_2

36. Which of the following compounds is bonded covalently?
(a) HCl (b) LiCl (c) NaCl (d) KCl

37. Which of the following compounds does not obey the octet rule with regard to its bonding?
    (a) NaCl (b) OF_2 (c) $AsCl_5$ (d) Br_2

38. Which of the following compounds has a double covalent bond in its structure?
    (a) N_2 (b) C_2H_2 (c) CH_4 (d) C_2H_4

39. Which of the following compounds does not obey the octet rule with regard to its bonding?
    (a) $CaCl_2$ (b) H_2Se (c) BF_3 (d) I_2

40. Which of the following <u>molecules</u> is <u>nonpolar</u>?

(a) 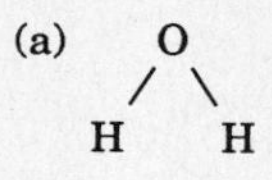

(b)
N
/ | \
H H H

(c) 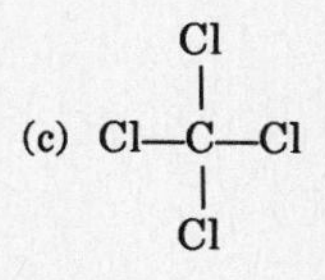

(d) 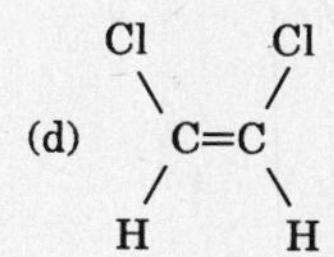

CHAPTER 7 Quiz Answers

1. (a) Cs· (b) Ṁg· (c) ·S̈: (d) ·Ȧl·

2. (a) Polar (b) Polar (c) Polar (d) Nonpolar

3. (a) Nonpolar (b) Polar (c) Polar (d) Nonpolar

4. (a) +7 (b) +4 (c) +7 (d) +5

5. O=O is least polar bond, then comes N—O, and C—O is most polar bond.

6. In an ionic bond there is a transfer of electrons between the atoms that form the bond. In a covalent bond there is a sharing of electrons between the atoms that form the bond.

7. The diatomic elements are hydrogen, oxygen, nitrogen, chlorine, bromine, iodine, and fluorine.

8. two

9. (a) H, Cl, Br, I, and F (b) O (c) N

10. (a) H:C:::N̈ or H—C≡N̈

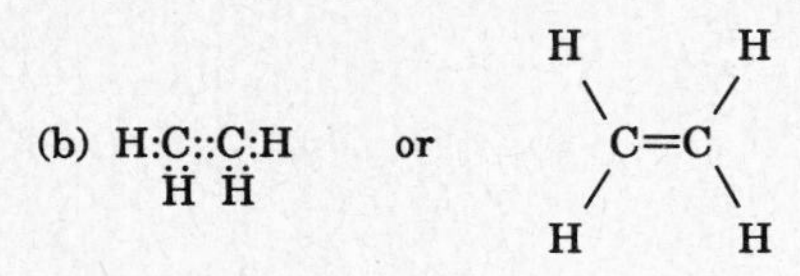

11. (a) PCl_5:

```
   Cl  Cl
    \ /
     P
   / | \
 Cl  |  Cl
     Cl
```

(b) BF_3:

```
        F
       /
  F—B
       \
        F
```

12.

```
     O
     ↑
  O←Br—O—H
     ↓
     O
```

15.

```
 :Ö ← C̈l—Ö—H
        ↓
       :Ö:
```

13. (a) Ionic (b) Covalent
(c) Ionic (d) Covalent

14. H_2Te, H_2Se, H_2S, H_2O
most covalent — least covalent

16. HF

17. (a) MgO (b) Al_2O_3 (c) Cs_2S

18. (a) 5+ (b) 5+ (c) 2+ (d) 6+

19. (a) The bonds are polar; the molecule is nonpolar.
(b) The bonds are nonpolar; the molecule is nonpolar.
(c) The bonds are polar; the molecule is polar.
(d) The bonds are polar; the molecule is polar.

20. The C—S bond is the most covalent.

21. c	22. a	23. b	24. d	25. b
26. a	27. a	28. c	29. b	30. b
31. c	32. a	33. a	34. c	35. d
36. a	37. c	38. d	39. c	40. c

CHAPTER 8 Quiz

1. What is the common name and the systematic name of solid CO_2?

2. Write the number that corresponds to each of the following prefixes:
(a) octa- (b) tri-
(c) tetra- (d) nona-
(e) penta- (f) hepta-

3. Write the formulas of the following binary compounds composed of two nonmetals:
(a) diphosphorus pentasulfide
(b) chlorine dioxide
(c) dinitrogen tetroxide
(d) dichlorine heptoxide

4. Write the formulas of the following binary compounds composed of a metal with a fixed oxidation number and a nonmetal:
(a) lithium iodide (b) calcium oxide
(c) strontium bromide (d) potassium phosphide
(e) rubidium sulfide (f) barium nitride

5. Explain the difference between:
(a) ferrous and ferric (b) cuprous and cupric
(c) cobaltous and cobaltic (d) stannous and stannic

6. Write the formulas of the following binary compounds composed of a metal with a variable oxidation number and a nonmetal:
(a) uranium(VI) fluoride (b) iron(III) nitride
(c) manganous chloride (d) ferrous sulfide
(e) cuprous oxide (f) mercurous chloride

7. Write the formulas of the following polyatomic ions:
(a) ammonium ion (b) hydroxide ion
(c) borate ion (d) hydrogen carbonate ion
(e) oxalate ion (f) sulfite ion

8. Write the formulas of the following ternary and higher compounds:
(a) cesium hydroxide (b) copper(I) arsenate
(c) ammonium sulfate (d) potassium carbonate
(e) ferric cyanide (f) cuprous sulfate

9. Write the names of the following binary compounds composed of two nonmetals:
(a) N_2O (b) NO_2 (c) SO_3 (d) N_2O_5

10. Write the names of the following binary compounds containing metals with fixed oxidation numbers:
(a) Cs_2O (b) Al_2S_3 (c) BaI_2
(d) $GaCl_3$ (e) K_2O (f) MgS

11. Find the oxidation number of the underlined element:
(a) $H\underline{Cl}O$ (b) $H\underline{Cl}O_2$ (c) $H\underline{Cl}O_3$ (d) $H\underline{Cl}O_4$

12. Write the names of the following binary compounds containing metals with variable oxidation numbers:
(a) CuS (b) $AuBr_3$ (c) FeO
(d) Cu_3P_2 (e) V_2O_5 (f) MnO_2

13. Write the names of the following ternary compounds:
(a) $CaCO_3$ (b) $NaNO_2$
(c) $NaOH$ (d) $Mg(OH)_2$
(e) $K_2Cr_2O_7$ (f) NH_4I

14. Write the formulas of the following acids:
(a) acetic acid (b) sulfurous acid
(c) chloric acid (d) hydrofluoric acid
(e) phosphoric acid (f) nitrous acid

15. Name the following compounds as inorganic acids:
(a) HIO_4 (b) $HClO_3$ (c) HBr
(d) HF (e) $HClO_2$ (f) HIO

16. Write the formulas of each of the following compounds designated by their common names:
(a) Epsom salts (b) muriatic acid
(c) marble (d) table salt

17. Write the common name of the compound designated by each formula:
(a) CaO (b) $CaCO_3$
(c) $CaSO_4 \cdot 2H_2O$ (d) H_2SO_4

18. Write the formula of the compound formed from each of the following pairs of ions:
(a) Al^{3+} and N^{3-} (b) V^{5+} and O^{2-}
(c) Fe^{3+} and $(OH)^{1-}$ (d) $(NH_4)^{1+}$ and S^{2-}

19. Write the formulas of the following compounds:
(a) tin(II) ion plus sulfide ion
(b) copper(I) ion plus phosphate ion
(c) iron(III) ion plus nitrite ion
(d) magnesium ion plus acetate ion

20. Write the correct name of each of the following compounds:
(a) $KClO_2$ (b) $Fe(CN)_3$
(c) P_2S_5 (d) CCl_4
(e) $HC_2H_3O_2$ (in water) (f) SnO_2

21. The formula for iron(III) oxide is:
(a) FeO (b) Fe_2O (c) FeO_3 (d) Fe_2O_3

22. The name of the compound $(NH_4)_2SO_4$ is:
(a) nitrogen sulfide (b) ammonium sulfate
(c) nitrogen sulfite (d) ammonium sulfide

23. The formula of copper(II) sulfate is:
(a) $Cu(SO_4)_2$ (b) $CuSO_4$
(c) Cu_2SO_4 (d) $CuSO_3$

24. The name of the compound $Fe(OH)_3$ is:
(a) iron(I) hydroxide (b) iron(III) hydroxide
(c) iron(III) oxide (d) iron(I) oxide

25. The formula for copper(I) oxide is:
(a) CuO (b) Cu_2O (c) CuO_2 (d) Cu_2O_3

26. The name of the compound $Hg(NO_3)_2$ is:
(a) mercury (II) nitrate
(b) mercury(II) nitrite
(c) mercury(I) nitrate
(d) mercury(II) nitrite

27. The formula of copper(II) phosphate is:
(a) $Cu(PO_4)_2$
(b) Cu_2PO_4
(c) $Cu_3(PO_4)_2$
(d) $Cu_2(PO_4)_3$

28. The name of the compound OsO_4 is:
(a) osmium(VIII) oxide
(b) osmium(IV) oxide
(c) osmium(II) oxide
(d) osmium oxide

29. The oxidation number of the Fe in FeO is:
(a) 1+
(b) 2+
(c) 3+
(d) 3–

30. The formula for copper(I) sulfite is:
(a) Cu_2S
(b) CuS
(c) Cu_2SO_4
(d) Cu_2SO_3

31. The name of the compound $Fe_2(CrO_4)_3$ is:
(a) iron(II) chromate
(b) iron(III) chromate
(c) iron chromium oxide
(d) iron chromic oxide

32. The formula for copper(I) sulfide is:
(a) Cu_2S
(b) CuS
(c) Cu_2SO_4
(d) Cu_2SO_3

33. The name of the compound $Al_2(CrO_4)_3$ is:
(a) aluminum chromate
(b) aluminum chromite
(c) aluminum chromium oxide
(d) aluminum chromic oxide

34. In the compound MgS, the magnesium atom has:
(a) gained one electron
(b) lost one electron
(c) gained two electrons
(d) lost two electrons

35. The formula for iron(III) nitrite is:
(a) FeN
(b) $Fe(NO_3)_3$
(c) $Fe(NO_2)_3$
(d) Fe_3NO_3

36. The name of the compound $Al_2(SO_3)_3$ is:
(a) aluminum sulfide
(b) aluminum sulfite
(c) aluminum sulfate
(d) aluminum sulfur oxide

37. In the compound Ag_2O, each silver atom has:
(a) gained one electron
(b) lost one electron
(c) gained two electrons
(d) lost two electrons

38. The formula for nickel(II) phosphate is:
(a) $Ni_3(PO_4)_2$
(b) $NiPO_4$
(c) $Ni_2(PO_4)_3$
(d) $Ni(PO_4)_2$

39. The name of the compound $Au_2(SO_4)_3$ is:
(a) gold(II) sulfate
(b) gold(III) sulfate
(c) gold(II) sulfite
(d) gold(III) sulfite

40. In the compound MnO_2, the manganese atom has:
(a) gained two electrons (b) lost two electrons
(c) gained four electrons (d) lost four electrons

41. The formula for copper(II) acetate is:
(a) Cu_2SO_4 (b) $CuC_2H_3O_2$
(c) $Cu_2C_2H_3O_2$ (d) $Cu(C_2H_3O_2)_2$

42. The name of the compound $FeSO_4$ is:
(a) iron(II) sulfate (b) iron(III) sulfate
(c) iron(II) sulfite (d) iron(III) sulfite

43. The oxidation number of U in the compound UO_3 is:
(a) 3+ (b) 4+ (c) 5+ (d) 6+

44. The formula for diphosphorus pentoxide is:
(a) P_2O_5 (b) P_2O_6 (c) PO_5 (d) P_2O

45. The name of the compound MnO_2 is:
(a) manganese dioxide (b) dimanganese oxide
(c) magnesium oxide (d) magnesium dioxide

46. The oxidation number of Cu in the compound Cu_2S is:
(a) 4+ (b) 3+ (c) 2+ (d) 1+

CHAPTER 8 Quiz Answers

1. The common name of CO_2 is dry ice. The systematic name is carbon dioxide.

2. (a) octa is 8 (b) tri is 3
(c) tetra is 4 (d) nona is 9
(e) penta is 5 (f) hepta is 7

3. (a) P_2O_5 (b) ClO_2 (c) N_2O_4 (d) Cl_2O_7

4. (a) LiI (b) CaO (c) $SrBr_2$
(d) K_3P (e) Rb_2S (f) Ba_3N_2

5. (a) ferrous is Fe^{2+} and ferric is Fe^{3+}
(b) cuprous is Cu^{1+} and cupric is Cu^{2+}
(c) cobaltous is Co^{2+} and cobaltic is Co^{3+}
(d) stannous is Sn^{2+} and stannic is Sn^{4+}

6. (a) UF_6 (b) FeN (c) $MnCl_2$
(d) FeS (e) Cu_2O (f) Hg_2Cl_2

7. (a) $(NH_4)^{1+}$ (b) $(OH)^{1-}$ (c) $(BO_3)^{3-}$
(d) $(HCO_3)^{1-}$ (e) $(C_2O_4)^{2-}$ (f) $(SO_3)^{2-}$

8. (a) CsOH (b) Cu_3AsO_4 (c) $(NH_4)_2SO_4$
(d) K_2CO_3 (e) $Fe(CN)_3$ (f) Cu_2SO_4

9. (a) dinitrogen monoxide (b) nitrogen dioxide
(c) sulfur trioxide (d) dinitrogen pentoxide

10. (a) cesium oxide (b) aluminum sulfide
(c) barium iodide (d) gallium chloride
(e) potassium oxide (f) magnesium sulfide

11. (a) 1+ (b) 3+ (c) 5+ (d) 7+

12. (a) copper(II) sulfide or cupric sulfide
(b) gold(III) bromide or auric bromide
(c) iron(II) oxide or ferrous oxide
(d) copper(II) phosphide or cupric phosphide
(e) vanadium(V) oxide
(f) manganese(IV) oxide

13. (a) calcium carbonate
(b) sodium nitrite
(c) sodium hydroxide
(d) magnesium hydroxide
(e) potassium dichromate
(f) ammonium iodide

14. (a) $HC_2H_3O_2$ (b) H_2SO_3 (c) $HClO_3$
(d) HF (e) H_3PO_4 (f) HNO_2

15. (a) periodic acid (b) chloric acid
(c) hydrobromic acid (d) hydrofluoric acid
(e) chlorous acid (f) hypoiodous acid

16. (a) Epsom salts is $MgSO_4 \cdot 7H_2O$
(b) muriatic acid is HCl
(c) marble is $CaCO_3$
(d) table salt is NaCl

17. (a) CaO is quicklime
(b) $CaCO_3$ is marble or limestone
(c) $CaSO_4 \cdot 2H_2O$ is gypsum
(d) H_2SO_4 is oil of vitriol

18. (a) AlN (b) V_2O_5 (c) $Fe(OH)_3$ (d) $(NH_4)_2S$

19. (a) SnS (b) Cu_3PO_4 (c) $Fe(NO_2)_3$ (d) $Mg(C_2H_3O_2)_2$

20. (a) potassium chlorite
(b) iron(III) cyanide or ferric cyanide
(c) diphosphorus pentasulfide
(d) carbon tetrachloride
(e) acetic acid
(f) tin(IV) oxide or stannic oxide

21. d 22. b 23. b 24. b 25. b

26. a 27. c 28. a 29. b 30. d

31. b 32. a 33. a 34. d 35. c

36. b 37. b 38. a 39. b 40. d

41. d 42. a 43. d 44. a 45. a

46. d

CHAPTER 9 Quiz

1. Find the molecular mass of the following compounds:
   (a) H_2SO_4 (b) $CaCO_3$ (c) $Ca(OH)_2$

2. Find the mass in grams of:
   (a) 2.00 moles of Na atoms
   (b) 0.500 mole of H_2O
   (c) 3.00×10^{23} molecules of CO_2
   (Hint: Use 6.02×10^{23} as Avogadro's number.)

3. Find the number of moles in:
   (a) 54.0 g of H_2O
   (b) 16.0 g of O_2
   (c) 3.00×10^{23} molecules of CO_2

4. Find the empirical formula of a compound with the following percent composition by mass:

   C = 30.0 percent H = 6.00 percent O = 64.0 percent

5. What is the percent composition of $CaCO_3$?

6. Determine the empirical formula of a compound that is 54.55% C, 9.09% H, and 36.36% O by mass. The molecular mass of this compound is 88. What is its molecular formula?

7. The compound fumaric acid has the empirical formula CHO. Its molecular mass is 116.0. What is its molecular formula?

8. How many grams of copper can be obtained from $1{,}6\overline{00}$ g of CuO?

9. Determine the molecular or formula mass of each of the following compounds:
   (a) $HC_2H_3O_2$
   (b) $(NH_4)_2SO_4$
   (c) $C_{12}H_{22}O_{11}$
   (d) $Fe_3(PO_4)_2$

10. Determine the moles of atoms in each element:
    (a) 10.0 g of Ca
    (b) $24\overline{0}$ g of C

11. Determine the moles of molecules in each compound:
    (a) $23\overline{0}$ g of NO_2
    (b) 1.70 g of H_2S

12. Determine the number of atoms of each element in Question 10.

13. Determine the number of molecules of each element in Question 11.

14. Determine the grams of each element from the number of moles of atoms:
    (a) 0.400 mole of Na atoms
    (b) 5.80 moles of I atoms

15. Determine the grams of each compound from the number of moles of molecules:
    (a) 8.30 moles of $HC_2H_3O_2$
    (b) 0.25 mole of $C_6H_{12}O_6$

16. What is the percent composition of Ag_2SO_4?

17. Find the empirical formula of a compound with the following percent composition by mass:

    N = 30.4 percent   O = 69.6 percent

18. Find the molecular formula of a compound whose empirical formula is CH_2N_2 and whose molecular weight is 126.

19. Determine the number of moles of each element in 940.0 g of $C_2H_4Br_2$.

20. Determine the number of grams of each element in Question 19.

21. How many grams are there in 5.00 moles of $CaCO_3$?
    (a) 5.00 g   (b) $5\bar{0}0$ g
    (c) $34\bar{0}$ g   (d) $1\bar{0}0$ g

22. How many moles are there in 68.4 g of $Al_2(SO_4)_3$?
    (a) 0.200 mole   (b) 1.00 mole
    (c) 2.00 moles   (d) 5.00 moles

23. How many grams are there in 0.400 mole of SO_2?
    (a) 25.6 g   (b) $16\bar{0}$ g
    (c) 19.2 g   (d) $12\bar{0}$ g

24. How many moles are there in 16.4 g of $Ca(NO_3)_2$?
    (a) 0.100 mole   (b) 10.0 moles
    (c) 5.00 moles   (d) 1.00 mole

25. How many grams are there in 0.250 mole of $C_6H_{12}O_6$?
    (a) 45.0 g   (b) 90.0 g
    (c) $18\bar{0}$ g   (d) $72\bar{0}$ g

26. How many moles are there in 63.0 g of $(NH_4)_2Cr_2O_7$?
    (a) 4.00 moles   (b) 1.00 mole
    (c) 0.500 mole   (d) 0.250 mole

27. The number of moles of carbon atoms in 5 moles of glucose ($C_6H_{12}O_6$) is:
    (a) 5 moles   (b) 20 moles
    (c) 30 moles   (d) 1 mole

28. How many grams are there in 0.200 mole of CO_2?
    (a) 88.0 g   (b) 8.80 g
    (c) 44.0 g   (d) 4.40 g

29. How many moles are there in 94.8 g of $Ca(C_2H_3O_2)_2$?
    (a) 0.600 mole   (b) 0.100 mole
    (c) 6.00 moles   (d) 1.00 mole

30. The number of moles of hydrogen in 15.0 moles of C_2H_6O is:
(a) 1.00 mole (b) 6.00 moles
(c) 10.0 moles (d) 90.0 moles

31. How many grams are there in 0.05 mole of pentane (C_5H_{12})?
(a) 3.6 g (b) 36 g
(c) 0.050 g (d) 1440 g

32. The number of moles of carbon in 8.00 moles of $C_{12}H_{22}O_{11}$ is:
(a) 8.00 moles (b) 88.0 moles
(c) 10.0 moles (d) 96.0 moles

33. How many grams are there in 28.0 moles of ammonia (NH_3)?
(a) 1.65 g (b) $42\bar{0}$ g
(c) 476 g (d) 392 g

34. The number of moles of hydrogen in 3.00 moles of CH_4 is:
(a) 3.00 moles (b) 6.00 moles
(c) 9.00 moles (d) 12.0 moles

35. How many grams are there in 0.500 mole of carbon dioxide (CO_2)?
(a) 44.0 g (b) 22.0 g
(c) 88.0 g (d) 11.0 g

36. The number of moles of chlorine in 8.00 moles of CCl_4 is:
(a) 1.00 mole (b) 4.00 moles
(c) 8.00 moles (d) 32.0 moles

37. How many grams are there in 0.500 mole of ammonium carbonate, $(NH_4)_2CO_3$?
(a) 96.0 g (b) 192 g
(c) 48.0 g (d) 24.0 g

38. The number of moles of chlorine atoms in 7.00 moles of $CHCl_3$ is:
(a) 1.00 mole (b) 7.00 moles
(c) 21.0 moles (d) 28.0 moles

39. How many grams are there in 0.250 mole of CH_4?
(a) 4.00 g (b) 8.00 g
(c) 12.0 g (d) 16.0 g

40. The number of moles of hydrogen atoms in 2.50 moles of H_3PO_4 is:
(a) 3.00 moles (b) 5.00 moles
(c) 7.50 moles (d) 10.0 moles

41. The percent by mass of S in the compound SO_2 is:
(a) 50% (b) 25% (c) 32% (d) 64%

CHAPTER 9 Quiz Answers

1. (a) 98.1 (b) 100.1 (c) 74.1

2. (a) 46.0 g (b) 9.00 g (c) 21.9 g

3. (a) 3.00 moles (b) 0.500 moles (c) 0.499 mole

4. $C_5H_{12}O_8$

5. Ca = 40.1 percent C = 12.0 percent O = 48.0 percent

6. In 100 g of compound there are 54.55 g C, 9.09 g H, and 36.36 g O.

$$? \text{ moles C atoms} = 54.55 \text{ g C} \times \frac{1 \text{ mole C atoms}}{12.0 \text{ g C}} = 4.55 \text{ moles}$$

$$? \text{ moles H atoms} = 9.09 \text{ g H} \times \frac{1 \text{ mole H atoms}}{1.0 \text{ g H}} = 9.00 \text{ moles}$$

$$? \text{ moles O atoms} = 36.36 \text{ g O} \times \frac{1 \text{ mole O atoms}}{16.00 \text{ g O}} = 2.273 \text{ moles}$$

The empirical formula is C_2H_4O and the molecular formula is $C_4H_8O_2$.

7. The formula mass of CHO = 29.0. 116.0/29.0 = 4. Therefore the molecular formula is $(CHO)_4$ or $C_4H_4O_4$.

8. The formula mass of CuO is 79.5.

$$\text{Percent Cu} = \frac{63.5}{79.5} \times 100 = 79.9$$

$$? \text{ grams Cu} = 16\overline{0}0 \text{ g CuO} \times 0.799 = 1.28 \times 10^3 \text{ g Cu}$$

9. (a) 60.0 (b) 132.1 (c) 342.0 (d) 357.4

10. (a) 0.249 mole (b) 20.0 moles

11. (a) 5.00 moles (b) 0.0499 mole

12. (a) 1.50×10^{23} (b) 1.20×10^{25}

13. (a) 3.01×10^{24} (b) 3.00×10^{22}

14. (a) 9.20 g (b) 736 g

15. (a) 498 g (b) 45 g

16. Ag = 69.2 percent, S = 10.3 percent, O = 20.5 percent

17. NO_2

18. $C_3H_6N_6$

19. 10.0 moles of C, 20.0 moles of H, 10.0 moles of Br

20. $12\overline{0}$ g of C, 20.0 g of H, $8\overline{00}$ g of Br

21. b	22. a	23. a	24. a	25. a
26. d	27. c	28. b	29. a	30. d
31. a	32. d	33. c	34. d	35. b

36. d 37. c 38. c 39. a 40. c

41. a

CHAPTER 10 Quiz

1. Write balanced chemical equations for the following reactions:

   (a) Sodium hydroxide + sulfuric acid → sodium sulfate + water

   (b) Sodium + water → sodium hydroxide + hydrogen gas

   (c) Potassium chloride $\overset{Heat}{\rightarrow}$ potassium chloride + oxygen gas

   (d) Sulfur dioxide + calcium oxide → calcium sulfite

   (e) Copper + silver nitrate → copper(II) nitrate + silver

2. Complete and balance the following reactions:

   (a) $Ca(OH)_2 + HCl \rightarrow$
   (b) $Zn + HCl \rightarrow$

   (c) $H_2SO_4 + NH_4OH \rightarrow$
   (d) $H_2 + Cl_2 \rightarrow$

   (e) $CaCO_3 \overset{Heat}{\rightarrow}$

3. Predict the products of the following reactions. If the reaction does not occur, write No reaction. (Check the activity series and solubility table.)

   (a) $Zn + H_2SO_4 \rightarrow$
   (b) $Cu + NiCl_2 \rightarrow$

   (c) $AgNO_3 + K_3AsO_4 \rightarrow$
   (d) $KCl + NaNO_3 \rightarrow$

4. For the reaction $5Zn + V_2O_5 \rightarrow 5ZnO + 2V$, list:
   (a) The oxidizing agent
   (b) The reducing agent
   (c) The element oxidized
   (d) The element reduced

5. Using an acid-base reaction, write a balanced equation for a reaction that produces potassium sulfate.

6. Write balanced chemical equations for each of the following reactions:

   (a) Calcium hydroxide + phosphoric acid → calcium phosphate + water

   (b) Lithium + oxygen gas → lithium oxide

   (c) Barium chlorate → barium chloride + oxygen gas

   (d) Nitrogen gas + hydrogen gas → ammonia

   (e) Silver nitrate + barium chloride → barium nitrate + silver chloride

7. Identify each of the reactions in Question 6 as a combination, decomposition, single-replacement or double-replacement reaction.

8. Predict the products and balance the equations of the following combination reactions:
   (a) $H_2 + I_2 \rightarrow$
   (b) $BaO + H_2O \rightarrow$
   (c) $K + Cl_2 \rightarrow$
   (d) $Na_2O + H_2O \rightarrow$

9. Predict the products and balance the equations of the following decomposition reactions:
   (a) $MgCO_3 \rightarrow$
   (b) $KClO_3 \rightarrow$
   (c) $Mg_3N_2 \rightarrow$
   (d) $LiOH \rightarrow$

10. Predict the products and balance the equations of the following single-replacement reactions. Use the activity series.
   (a) $Zn + HCl \rightarrow$
   (b) $Cu + HCl \rightarrow$
   (c) $Zn + Pb(NO_3)_2 \rightarrow$
   (d) $Ni + Al(NO_3)_3 \rightarrow$

11. Predict the products and balance the equations of the following double-replacement reactions. Use the solubility table.
   (a) $AgNO_3 + H_2S \rightarrow$
   (b) $BaCl_2 + (NH_4)_2CO_3 \rightarrow$
   (c) $KNO_3 + BaCl_2 \rightarrow$
   (d) $H_2SO_4 + NH_4OH \rightarrow$

12. Write an acid-base reaction that will produce each of the following salts:
   (a) $CaSO_4$
   (b) $Fe(NO_3)_3$

13. Write a single-replacement reaction that will produce each of the following salts:
   (a) $MgCl_2$
   (b) $Al_2(SO_4)_3$

14. Complete the following statements:

   (a) The loss of electrons is ________________.

   (b) The gain of electrons is ________________.

   (c) A substance that causes something else to be oxidized is called a(n) ________________.

   (d) A substance that causes something else to be reduced is called a(n) ________________.

15. For each of the following reactions, determine the substance oxidized and the substance reduced.
   (a) $CuO + H_2 \rightarrow Cu + H_2O$
   (b) $4Al + 3O_2 \rightarrow 2Al_2O_3$

16. For each equation in Question 15, identify the oxidizing agent and the reducing agent.

17. A student is asked to balance the following equation:

   Magnesium hydroxide + hydrochloric acid → magnesium chloride + water

   The student works the problem and writes the following solution:

   $MgOH + HCl \rightarrow MgCl + H_2O$

   The attempt is marked incorrect. Explain what is wrong with the answer and write the correct balanced equation.

18. Write a balanced equation for a reaction that produces strontium chloride.

19. Write a balanced equation for a reaction that produces carbon dioxide as one of the products.

20. Consider an experiment in which metal A is immersed in a water solution of ions of metal B. For each of the following pairs of A and B, determine whether a reaction occurs. If a reaction does occur, write the balanced equation, assuming that the B ions are part of nitrate salts.

	A	B		A	B
(a)	Cu	Zn	(b)	Mg	Zn
(c)	Zn	Cu	(d)	Cu	Ni

21. Balance the following equation:

$\_\_\_\_HCl + \_\_\_\_Ca(OH)_2 \rightarrow \_\_\_\_CaCl_2 + \_\_\_\_H_2O$

The sum of the coefficients is:
(a) 4 (b) 5 (c) 6 (d) 7

22. In the following reaction:

$Mg + H_2SO_4 \rightarrow MgSO_4 + H_2$

the substance being oxidized is:
(a) magnesium (b) hydrogen ion
(c) sulfur (d) oxygen

23. In the reaction in Question 22, the substance being reduced is:
(a) magnesium (b) hydrogen ion
(c) sulfur (d) oxygen

24. The reaction in Question 22 can be classified as a:
(a) combination reaction
(b) decomposition reaction
(c) single-replacement reaction
(d) double-replacement reaction

25. Balance the following equation:

$\_\_\_\_C_2H_6 + \_\_\_\_O_2 \rightarrow \_\_\_\_CO_2 + \_\_\_\_H_2O$

The sum of the coefficients is:
(a) 10 (b) 9 (c) 4 (d) 19

26. Which substance is being oxidized in the following equation?

$Zn + CuSO_4 \rightarrow ZnSO_4 + Cu$

(a) zinc metal (b) copper(II) ion
(c) sulfate ion (d) copper metal

27. For the reaction in Question 26, the substance being reduced is:
(a) zinc (b) copper(II) ion
(c) sulfate ion (d) copper metal

28. The reaction $Ca(s) + S(s) \rightarrow CaS(s)$ can be classified as:
(a) single-replacement reaction
(b) double-replacement reaction
(c) combination reaction
(d) decomposition reaction

29. Balance the following equation:

    ____H_2SO_4 + ____$Al(OH)_3$ → ____$Al_2(SO_4)_3$ + ____H_2O

    The sum of the coefficients is:
    (a) 4 (b) 9 (c) 12 (d) 15

30. In the following reaction:

    $Zn + H_2SO_4 \rightarrow ZnSO_4 + H_2$

    the substance being oxidized is:
    (a) zinc (b) hydrogen ion
    (c) sulfur (d) oxygen

31. In the reaction in Question 30, the substance being reduced is:
    (a) zinc (b) hydrogen ion
    (c) sulfur (d) oxygen

32. The reaction in Question 29 can be classified as a:
    (a) combination reaction
    (b) decomposition reaction
    (c) single-replacement reaction
    (d) double-replacement reaction

33. Balance the following equation:

    ____$KClO_3$ → ____KCl + ____O_2

    The sum of the coefficients is:
    (a) 6 (b) 7 (c) 5 (d) 4

34. Which substance is being oxidized in the following equation?

    $H_2 + CuO \rightarrow H_2O + Cu$

    (a) hydrogen gas (b) copper(II) ion
    (c) oxide ion (d) copper metal

35. For the reaction in Question 34, the substance being reduced is:
    (a) hydrogen gas (b) copper(II) ion
    (c) oxide ion (d) copper metal

36. The reaction in Question 34 can be classified as:
    (a) single-replacement reaction
    (b) double-replacement reaction
    (c) combination reaction
    (d) decomposition reaction

37. Balance the following equation:

    ____HBr + ____$Fe(OH)_3$ → ____$FeBr_3$ + ____H_2O

    The sum of the coefficients is:
    (a) 6 (b) 7 (c) 8 (d) 9

38. In the following reaction:

$Cl_2 + 2NaBr \rightarrow 2NaCl + Br_2$

the substance being oxidized is:
(a) chlorine (b) sodium ion
(c) bromide ion (d) sodium chloride

39. In the reaction in Question 38, the substance being reduced is:
(a) chlorine (b) sodium ion
(c) bromide ion (d) sodium chloride

40. The reaction in Question 38 can be classified as a:
(a) combination reaction
(b) decomposition reaction
(c) single-replacement reaction
(d) double-replacement reaction

41. Balance the following equation:

____ Fe_2O_3 + ____ $CO \rightarrow$ ____ CO_2 + ____ Fe

The sum of the coefficients is:
(a) 6 (b) 7 (c) 8 (d) 9

42. In the reaction in Question 41, the atom being oxidized is:
(a) iron(III) ion (b) oxide ion
(c) carbon (d) iron metal

43. In the reaction in Question 41, the substance being reduced is:
(a) iron(III) ion (b) oxide ion
(c) carbon (d) iron metal

44. The following reaction:

$AgNO_3 + HCl \rightarrow AgCl + HNO_3$

can be classified as a:
(a) combination reaction
(b) decomposition reaction
(c) single-replacement reaction
(d) double-replacement reaction

45. Balance the following equation:

____ K + ____ $H_2O \rightarrow$ ____ KOH + ____ H_2

The sum of the coefficients is:
(a) 6 (b) 7 (c) 8 (d) 9

46. In the reaction in Question 45, the substance being oxidized is:
(a) potassium metal (b) oxide ion
(c) hydrogen ion (d) nothing is oxidized

47. In the reaction in Question 45, the substance being reduced is:
(a) potassium metal (b) oxide ion
(c) hydrogen ion (d) nothing is reduced

48. The following reaction:

    $2KClO_3 \rightarrow 2KCl + 3O_2$

    can be classified as a:
    (a) combination reaction
    (b) decomposition reaction
    (c) single-replacement reaction
    (d) double-replacement reaction

49. Balance the following equation:

    ____$Ca(ClO_3)_2 \rightarrow$ ____$CaCl_2 +$ ____O_2

    The sum of the coefficients is:
    (a) 3 (b) 4 (c) 5 (d) 6

50. In the reaction in Question 49, the substance being oxidized is:
    (a) calcium (b) oxygen
    (c) chlorine (d) nothing is oxidized

51. In the reaction in Question 49, the atom being reduced is:
    (a) calcium (b) oxygen
    (c) chlorine (d) nothing is reduced

52. The reaction in Question 49 can be classified as a:
    (a) combination reaction
    (b) decomposition reaction
    (c) single-replacement reaction
    (d) double-replacement reaction

CHAPTER 10 Quiz Answers

1. (a) $2NaOH + H_2SO_4 \rightarrow Na_2SO_4 + 2H_2O$
   (b) $2Na + 2H_2O \rightarrow 2NaOH + H_2$
   (c) $2KClO_3 \rightarrow 2KCl + 3O_2$
   (d) $SO_2 + CaO \rightarrow CaSO_3$
   (e) $Cu + 2AgNO_3 \rightarrow Cu(NO_3)_2 + 2Ag$

2. (a) $Ca(OH)_2 + 2HCl \rightarrow CaCl_2 + 2H_2O$
   (b) $Zn + 2HCl \rightarrow ZnCl_2 + H_2$
   (c) $H_2SO_4 + 2NH_4OH \rightarrow (NH_4)_2SO_4 + 2H_2O$
   (d) $H_2 + Cl_2 \rightarrow 2HCl$
   (e) $CaCO_3 \rightarrow CaO + CO_2$

3. (a) $Zn + H_2SO_4 \rightarrow ZnSO_4 + H_2$
   (b) $Cu + NiCl_2 \rightarrow$ No reaction
   (c) $3AgNO_3 + K_3AsO_4 \rightarrow Ag_3AsO_4 + 3KNO_3$
   (d) $KCl + NaNO_3 \rightarrow$ No reaction

4. (a) V_2O_5 is the oxidizing agent.
   (b) Zn is the reducing agent.
   (c) Zn is oxidized.
   (d) V is reduced.

5. (a) $2KOH + H_2SO_4 \rightarrow K_2SO_4 + 2H_2O$

6. (a) $3Ca(OH)_2 + 2H_3PO_4 \rightarrow Ca_3(PO_4)_2 + 6H_2O$
   (b) $4Li + O_2 \rightarrow 2Li_2O$
   (c) $Ba(ClO_3)_2 \rightarrow BaCl_2 + 3O_2$
   (d) $N_2 + 3H_2 \rightarrow 2NH_3$
   (e) $2AgNO_3 + BaCl_2 \rightarrow Ba(NO_3)_2 + 2AgCl$

7. (a) Double-replacement reaction
   (b) Combination reaction
   (c) Decomposition reaction
   (d) Combination reaction
   (e) Double-replacement reaction

8. (a) $H_2 + I_2 \rightarrow 2HI$
   (b) $BaO + H_2O \rightarrow Ba(OH)_2$
   (c) $2K + Cl_2 \rightarrow 2KCl$
   (d) $Na_2O + H_2O \rightarrow 2NaOH$

9. (a) $MgCO_3 \rightarrow MgO + CO_2$
   (b) $2KClO_3 \rightarrow 2KCl + 3O_2$
   (c) $Mg_3N_2 \rightarrow 3Mg + N_2$
   (d) $2LiOH \rightarrow Li_2O + H_2O$

10. (a) $Zn + 2HCl \rightarrow ZnCl_2 + H_2$
    (b) $Cu + HNO_3 \rightarrow$ No single replacement reaction
    (c) $Zn + Pb(NO_3)_2 \rightarrow Zn(NO_3)_2 + Pb$
    (d) $Ni + Al(NO_3)_3 \rightarrow$ No reaction

11. (a) $2AgNO_3 + H_2S \rightarrow Ag_2S + 2HNO_3$
    (b) $BaCl_2 + (NH_4)_2CO_3 \rightarrow BaCO_3 + 2NH_4Cl$
    (c) $KNO_3 + BaCl_2 \rightarrow$ No reaction
    (d) $H_2SO_4 + 2NH_4OH \rightarrow (NH_4)_2SO_4 + 2H_2O$

12. (a) $Ca(OH)_2 + H_2SO_4 \rightarrow CaSO_4 + 2H_2O$
    (b) $Fe(OH)_3 + 3HNO_3 \rightarrow Fe(NO_3)_3 + 3H_2O$

13. (a) $Mg + 2HCl \rightarrow MgCl_2 + H_2$
    (b) $2Al + 3H_2SO_4 \rightarrow Al_2(SO_4)_3 + 3H_2$

14. (a) oxidation
    (b) reduction
    (c) oxidizing agent
    (d) reducing agent

15. (a) The H_2 is oxidized and the Cu^{2+} is reduced.
    (b) The Al is oxidized and the O_2 is reduced.

16. (a) The H_2 is the reducing agent and the CuO is the oxidizing agent.
    (b) The Al is the reducing agent and the O_2 is the oxidizing agent.

17. Magnesium hydroxide has the formula $Mg(OH)_2$, and magnesium chloride has the formula $MgCl_2$. The correct balanced equation is

    $Mg(OH)_2 + 2HCl \rightarrow MgCl_2 + 2H_2O$.

18. $Sr(OH)_2 + 2HCl \rightarrow SrCl_2 + 2H_2O$

19. $CH_4 + 2O_2 \rightarrow CO_2 + 2H_2O$

20. (a) No reaction

    (b) $Mg + Zn(NO_3)_2 \rightarrow Mg(NO_3)_2 + Zn$

    (c) $Zn + Cu(NO_3)_2 \rightarrow Zn(NO_3)_2 + Cu$

    (d) No reaction

21. c	22. a	23. b	24. c	25. d
26. a	27. b	28. c	29. c	30. a
31. b	32. d	33. b	34. a	35. b
36. a	37. c	38. c	39. a	40. c
41. d	42. c	43. a	44. d	45. b
46. a	47. c	48. b	49. c	50. b
51. c	52. b			

CHAPTER 11 Quiz

1. Hydrogen peroxide (H_2O_2) decomposes to produce water and oxygen.
   (a) Write a balanced equation to describe this process.
   (b) When 68.0 g of H_2O_2 decompose, how many moles of O_2 are produced?
   (c) How many grams of H_2O are formed?

2. Consider the unbalanced equation $ZnS + O_2 \rightarrow ZnO + SO_2$.
   (a) Balance the equation.
   (b) When 0.50 mole of ZnS is reacted, how many grams of O_2 are used?
   (c) How many moles of ZnO and SO_2 are formed?

3. The burning of acetylene is expressed as follows:

   $C_2H_2 + O_2 \rightarrow CO_2 + H_2O$ (unbalanced)

   (a) Balance the equation.
   (b) Given that 22.0 g of CO_2 are formed, how many moles of O_2 are needed?
   (c) How many grams of H_2O are also formed?

4. Assume that 8.0 g of H_2 are reacted with 48 g of O_2.
   (a) Write the balanced equation for the reaction.
   (b) What is the limiting reagent in the reaction?
   (c) How many grams of H_2O are formed in this reaction?

5. Consider the unbalanced equation $H_2 + O_2 \rightarrow H_2O$.
   (a) Balance the equation.
   (b) How many moles of hydrogen gas and oxygen gas are needed to produce 30.0 moles of water?

6. Propane (C_3H_8) burns with oxygen gas to produce carbon dioxide and water.
   (a) Write a balanced equation for this reaction.
   (b) How many grams of oxygen are needed to burn 352.0 g of propane?
   (c) How many grams of carbon dioxide and water are produced?

7. Calcium hydroxide reacts with phosphoric acid to produce calcium phosphate and water.
   (a) Write a balanced equation for this reaction.
   (b) How many moles of calcium hydroxide are required to neutralize 8.00 moles of phosphoric acid?
   (c) How many moles of calcium phosphate and water are produced?

8. A 4.00 g sample containing a mixture of $KClO_3$ and NaCl is heated until all of the $KClO_3$ decomposes to form KCl and O_2. The oxygen escapes. After heating, the sample weighs 3.00 g. Find the percentage of $KClO_3$ and of NaCl in the original mixture.

9. How many grams of air are necessary for the combustion of 18.0 g of pentane (C_5H_{12})? (Hint: Assume that air contains 23.0 percent oxygen by mass.)

10. Glucose ($C_6H_{12}O_6$) reacts with oxygen to produce carbon dioxide and water.
    (a) Write a balanced equation for this reaction.
    (b) How many grams of oxygen are needed to react with 9.00 g of glucose?

11. Magnesium nitride (Mg_3N_2) reacts with water to produce magnesium hydroxide and ammonia gas.
    (a) Write a balanced equation for this reaction.
    (b) How many grams of water are needed to react with 5.00 g of Mg_3N_2?

12. How many grams of hydrogen gas can be produced by reacting 6.50 g of zinc with a sufficient amount of hydrochloric acid?

13. Silver nitrate reacts with copper to produce copper(II) nitrate plus silver metal.
    (a) Write a balanced equation for this reaction.
    (b) How many grams of copper are needed to produce 21.6 g of silver?

14. How many grams of water can be produced from 10.00 g of H_2 and 320.0 g of O_2?

15. How many moles of NaCl can be produced from 9.20 g of Na and 14.0 g of Cl_2?

16. How many grams of ammonia (NH_3) can be prepared from 30.0 g of H_2 and 112 g of N_2?

17. How many grams of CO_2 can be formed from 3.60 g of C and 12.80 g of O_2?

18. Consider the equation $Ca_3P_2 + H_2O \rightarrow PH_3 + Ca(OH)_2$.
    (a) Balance the equation.
    (c) How many grams of $Ca(OH)_2$ can be produced from 546 g of Ca_3P_2 and 72.0 g of H_2O?

19. How many moles of acetylene (C_2H_2) can be produced from 32.0 g of CaC_2 and 36.0 g of H_2O? (Hint: $Ca(OH)_2$ is also produced in this reaction.)

20. Elementary phosphorus can be produced via the reaction

    $Ca_3(PO_4)_2 + 3SiO_2 + 5C \rightarrow 3CaSiO_3 + 5CO + P_2$

    Determine how many moles of each product are formed from 10.0 moles of $Ca_3(PO_4)_2$, 21.0 moles of SiO_2, and 45.0 moles of C.

21. Given the following balanced equation:

    $H_2 + Cl_2 \rightarrow 2HCl$

    How many moles of HCl can be produced from 2.5 moles of H_2 and 2.5 moles of Cl_2?
    (a) 2.5 moles (b) 2.0 moles
    (c) 5.0 moles (d) 1.25 moles

22. How many moles of CO_2 are produced from the combustion of 0.25 mole of C_3H_8? The balanced equation for the reaction is:

    $C_3H_8 + 5O_2 \rightarrow 3CO_2 + 4H_2O$

    (a) 0.75 mole (b) 3.0 moles
    (c) 0.25 mole (d) 0.083 mole

23. Given the following balanced equation:

    $2H_2 + O_2 \rightarrow 2H_2O$

    How many moles of water can be produced from 10.0 moles of H_2 and 4.00 moles of O_2?
    (a) 10.0 moles (b) 8.00 moles
    (c) 4.00 moles (d) 14.0 moles

24. Given the following balanced equation:

    $2Al + 6HCl \rightarrow 2AlCl_3 + 3H_2$

    How many grams of aluminum chloride can be formed from 135 g of Al with excess hydrochloric acid?
    (a) 668 g (b) 334 g (c) 448 g (d) 167 g

25. Given the following balanced equation for the formation of acetylene from calcium carbide and water:

    $CaC_2 + 2H_2O \rightarrow Ca(OH)_2 + C_2H_2$

    How many grams of acetylene can be formed from 64 g of calcium carbide and 72 g of water?
    (a) 72 g (b) 36 g (c) 52 g (d) 26 g

26. If the final yield is 50.0% of the theoretical yield, how many grams of CS_2 are obtained from the reaction of 24.0 g of C and 96.0 g of S. The balanced equation is: $C + 2S \rightarrow CS_2$.
    (a) 114 g (b) 57.0 g (c) 152 g (d) 76.0 g

27. A student decomposes 12.26 g of $KClO_3$ according to the following equation: $2KClO_3 \rightarrow 2KCl + 3O_2$. If the student obtains 5.22 g of KCl, what is the student's percent yield?
(a) 90.0% (b) 50.0% (c) 70.0% (d) 35.5%

28. How many moles of ozone (O_3) may be formed from the reaction of 6.00 moles of oxygen gas (O_2)?
(a) 3.00 moles (b) 6.00 moles
(c) 2.00 moles (d) 4.00 moles

29. How many grams of NaCl can be formed from 4.60 g of Na and excess Cl_2?
(a) 23.4 g (b) 11.7 g (c) 46.8 g (d) 7.1 g

30. How many moles of Ag metal can be formed from the reaction of 34.0 g of $AgNO_3$ and excess zinc metal?
(a) 0.200 mole (b) 0.400 mole
(c) 0.800 mole (d) 2.00 moles

CHAPTER 11 Quiz Answers

1. (a) $2H_2O_2 \rightarrow 2H_2O + O_2$ (b) 1.00 mole of O_2 (c) 36.0 g of H_2O

2. (a) $2ZnS + 3O_2 \rightarrow 2ZnO + 2SO_2$ (b) 24 g of O_2
(c) 0.50 mole of ZnO and 0.50 mole of SO_2

3. (a) $2C_2H_2 + 5O_2 \rightarrow 4CO_2 + 2H_2O$
(b) 0.625 mole of O_2 (c) 4.50 g of H_2O

4. (a) $2H_2 + O_2 \rightarrow 2H_2O$ (b) Oxygen is limiting.
(c) 54 g of H_2O

5. (a) $2H_2 + O_2 \rightarrow 2H_2O$
(b) 30.0 moles of H_2 and 15.0 moles of O_2

6. (a) $C_3H_8 + 5O_2 \rightarrow 3CO_2 + 4H_2O$
(b) $128\overline{0}$ g of O_2
(c) 1056 g of CO_2 and 576.0 g of H_2O

7. (a) $3Ca(OH)_2 + 2H_3PO_4 \rightarrow Ca_3(PO_4)_2 + 6H_2O$
(b) 12.0 moles of $Ca(OH)_2$
(c) 4.00 moles of $Ca_3(PO_4)_2$ and 24.0 moles of H_2O

8. 63.8 percent $KClO_3$ and 36.2 percent NaCl

9. 278 g of air

10. (a) $C_6H_{12}O_6 + 6O_2 \rightarrow 6CO_2 + 6H_2O$
(b) 9.60 g of oxygen

11. (a) $Mg_3N_2 + 6H_2O \rightarrow 3Mg(OH)_2 + 2NH_3$
    (b) 5.40 g of water

12. 0.200 g of hydrogen gas (H_2)

13. (a) $Cu + 2AgNO_3 \rightarrow Cu(NO_3)_2 + 2Ag$
    (b) 6.40 g of copper

14. 180.0 g of water

15. 0.400 mole of NaCl

16. 136 g of NH_3

17. 13.2 g of CO2

18. (a) $Ca_3P_2 + 6H_2O \rightarrow 2PH_3 + 3Ca(OH)_2$
    (b) 148 g of $Ca(OH)_2$

19. 0.500 mole of C_2H_2

20. 21.0 moles of $CaSiO_3$, 35.0 moles of CO, and 7.0 moles of P_2

21. c	22. a	23. b	24. a	25. d
26. b	27. c	28. d	29. b	30. a

CHAPTER 12 Quiz

1. How many kcal of heat are produced upon the formation of 2.00 moles of $CO_2(g)$ from its elements?

   $C(\text{graphite}) + O_2(g) \rightarrow CO_2(g) + 94.1\ \text{kcal}$

2. Determine the ΔH_R for the following reaction:

   $2C_2H_6(g) + 7O_2(g) \rightarrow 4CO_2(g) + 6H_2O\ (\ell)$

3. Given the following equations, calculate the ΔH_f of $SO_2(g)$ formed from S(s) and $O_2(g)$.

   $SO_2(g) + \frac{1}{2}\ O_2(g) \rightarrow SO_3(g) + 23.49\ \text{kcal}$

   $S(s) + 1\frac{1}{2}\ O_2(g) \rightarrow SO_3(g) + 94.45\ \text{kcal}$

4. Match each term on the left with its definition on the right.

   (a) Exothermic reaction
   (b) Endothermic reaction
   (c) Heat of formation
   (d) Heat of reaction

   (1) The heat released or absorbed when a compound is formed from its elements
   (2) A reaction that releases heat to the environment
   (3) The change in heat content during the course of a reaction
   (4) A reaction that absorbs heat from the environment

5. Complete each sentence:

   (a) The symbol H, usually printed in italics, stands for the term ____________________.

   (b) The change in heat content during a chemical reaction is called the ____________________.

6. What are the standard-state conditions?

7. Match each term on the left with its definition on the right.

   (a) Calorie
   (b) Specific heat of water
   (c) Enthalpy

   (1) The unit that is a measure of the number of calories needed to raise the temperature of 1 gram of a substance by 1 Celsius degree
   (2) The energy associated with a particular substance
   (3) The amount of heat needed to raise 1 g of water 1°C

8. How many calories are needed to raise the temperature of 400.0 g of water from 20.0°C to 50.0°C?

9. Calculate the amount of heat needed to raise the temperature of 50.0 g of copper from 50.0°C to 70.0°C. (Hint: The specific heat of Cu is 0.0920 cal/g-°C.)

10. Suppose that 10.00 kcal of heat are added to a sample of gold that weighs 4.000 kg. What is the final temperature of the gold if its initial temperature is 20.0°C? (Hint: The specific heat of gold is 0.0308 cal/g-°C.)

11. Write a thermochemical equation for the formation of the following compounds from their elements:

    (a) $HCl(g)$ (b) $HI(g)$

    (Hint: ΔH_f° for $HCl(g)$ is –22.1 kcal/mole, and ΔH_f° for $HI(g)$ is +6.2 kcal/mole.)

12. Calculate the heat released when 170.0 g of $H_2S(g)$ are produced from its elements. (Hint: ΔH_f° for H_2S is –4.8 kcal/mole.)

13. Carbon dioxide is formed from its elements as follows:

    $C(graphite) + O_2(g) \rightarrow CO_2(g) + 94.1\ kcal$

    How many kilocalories of heat are produced when $22\overline{0}$ g of $CO_2(g)$ are formed?

14. Calculate the heat released when 12.8 g of $CH_4(g)$ are produced from its elements. (Hint: ΔH_f° of $CH_4(g)$ is –17.9 kcal/mole.)

15. Determine the ΔH_R for the following reaction:

    $2C_2H_2(g) + 5O_2(g) \rightarrow 4CO_2(g) + 2H_2O(\ell)$

    (Hint: ΔH_f° of $C_2H_2(g)$ = –20.2 kcal/mole, ΔH_f° of $CO_2(g)$ = –94.1 kcal/mole, and ΔH_f° of $H_2O(\ell)$ = –68.3 kcal/mole.)

16. Calculate the amount of heat released when 8.00 moles of $CO_2(g)$ are formed from the following reaction:

    $C(graphite) + 2H_2O(g) \rightarrow CO_2(g) + 2H_2(g) - 31.39$ kcal

17. Calculate the ΔH_f° of CuO(s) from the following information:

    $3CuO(s) + 2NH_3(g) \rightarrow 3Cu(s) + 3H_2O(\ell) + N_2(g) + 71.1$ kcal

    Also note that ΔH_f° of $NH_3(g)$ is –11.04 kcal/mole and that ΔH_f° of $H_2O(\ell)$ is –68.3 kcal/mole.

18. The ΔH_R for the burning of carbon is 94.0 kcal/mole. The coal used in a coal-fired power plant is 70.0 percent carbon. How much energy does 10.00 kg of coal yield?

19. Given that ΔH_f° for $H_2O(\ell)$ is –68.3 kcal/mole and that ΔH_f° for $H_2O(g)$ is –57.8 kcal/mole, calculate the heat needed to produce the following change in 5.00 moles of water: $H_2O(\ell) \rightarrow H_2O(g)$.

20. In a simple calorimeter, a reaction is carried out in a vessel submerged in water. After the reaction occurs, $42\bar{0}$ g of water increase in temperature from 25.0°C to 40.0°C. What is the amount of energy released by the reaction?

21. The number of calories needed to raise the temperature of 50 g of water from 20°C to 40°C is:
    (a) 100 cal (b) 200 cal
    (c) 500 cal (d) 1,000 cal

22. The number of calories needed to raise the temperature of 200 g of water from 20°C to 50°C is:
    (a) 6,000 cal (b) 4,000 cal
    (c) 10,000 cal (d) 14,000 cal

23. The number of kcal needed to raise the temperature of 500 g of water from 20°C to 60°C is:
    (a) 5 kcal (b) 10 kcal
    (c) 20 kcal (d) 40 kcal

24. If 8,000 cal of heat are added to 200 g of water at 20°C, what is the final temperature of the water?
    (a) 60°C (b) 40°C (c) 20°C (d) 80°C

25. If 1.000 cal = 4.184 J, then how many joules is 41.840 cal?
    (a) 10 J (b) 100 J (c) 175.1 J (d) 0.1 J

26. Calculate the ΔH for the reaction

    $CH_4(g) + 2O_2(g) \rightarrow CO_2(g) + 2H_2O(\ell)$

    given that:

    $\Delta H(CH_4) = -17.9$ kcal/mole
    $\Delta H(O_2) = 0.0$ kcal/mole
    $\Delta H(CO_2) = -94.1$ kcal/mole
    $\Delta H(H_2O) = -68.3$ kcal/mole

    (a) $\Delta H = -144.5$ kcal (b) $\Delta H = -248.6$ kcal
    (c) $\Delta H = -306.9$ kcal (d) $\Delta H = -212.8$ kcal

27. How much energy is released when 72.0 g of water reacts with sulfur trioxide in the following reaction?

    $SO_3(g) + H_2O(\ell) \rightarrow H_2SO_4 + 31.14$ kcal

    (a) 31.14 kcal
    (b) 62.28 kcal
    (c) 93.42 kcal
    (d) 124.6 kcal

28. Calculate the ΔH for the reaction

    $2C_2H_6(g) + 7O_2(g) \rightarrow 4CO_2(g) + 6H_2O(\ell)$

    given that:

    $\Delta H(C_2H_6) = -20.2$ kcal/mole
    $\Delta H(O_2) = 0.0$ kcal/mole
    $\Delta H(CO_2) = -94.1$ kcal/mole
    $\Delta H(H_2O) = -68.3$ kcal/mole

    (a) $\Delta H = -826.6$ kcal
    (b) $\Delta H = -745.8$ kcal
    (c) $\Delta H = -786.2$ kcal
    (d) $\Delta H = -182.6$ kcal

29. How much energy is released when 11.0 g of carbon dioxide is formed from its elements?

    $C(s) + O_2(g) \rightarrow CO_2(g) + 94.1$ kcal

    (a) 94.1 kcal
    (b) 47.1 kcal
    (c) 23.5 kcal
    (d) 188.2 kcal

30. Calculate the ΔH for the reaction

    $C_6H_{12}O_6(s) + 6O_2(g) \rightarrow 6CO_2(g) + 6H_2O(\ell)$

    given that:

    $\Delta H(C_6H_{12}O_6) = -304.4$ kcal/mole
    $\Delta H(O_2) = 0.0$ kcal/mole
    $\Delta H(CO_2) = -94.1$ kcal/mole
    $\Delta H(H_2O) = -68.3$ kcal/mole

    (a) $\Delta H = -1278.8$ kcal
    (b) $\Delta H = -974.4$ kcal
    (c) $\Delta H = -304.4$ kcal
    (d) $\Delta H = -670.0$ kcal

31. How much energy is absorbed from the environment when 32.0 g of hydrogen iodide, HI(g) is formed from its elements?

    $\frac{1}{2}H_2(g) + \frac{1}{2}I_2(g) \rightarrow HI(g) - 6.20$ kcal

    (a) 6.20 kcal
    (b) 1.55 kcal
    (c) 3.10 kcal
    (d) 6.20 kcal

32. The change in heat content of a substance during a chemical reaction is known as the:
    (a) heat of reaction
    (b) heat of formation
    (c) power
    (d) calories

33. How much energy is absorbed from the environment when 12.8 g of sulfur dioxide, $SO_2(g)$ is formed from its elements?

$S(s) + O_2(g) \rightarrow SO_2(g) - 71.0$ kcal

(a) –71.0 kcal
(b) –142 kcal
(c) –7.1 kcal
(d) –14.2 kcal

34. Choose the correct statement from the choices below:
(a) a joule is a larger unit of energy than a calorie
(b) a calorie is a larger unit of energy than a joule
(c) a calorie is a larger unit of energy than a kcal
(d) the unit of specific heat is "calories."

35. In an exothermic reaction the sum of the heat content of the reactants is ________________ the sum of the heat content of the products.
(a) greater than
(b) less than
(c) equal to
(d) sometimes greater and sometimes less than

CHAPTER 12 Quiz Answers

1. 188 kcal

2. –745.8 kcal

3. 70.96 kcal

4. (1) c (2) a (3) d (4) b

5. (a) enthalpy (b) heat of reaction

6. Standard-state conditions are 25°C and 1 atm.

7. (1) b (2) c (3) a

8. 12,$\overline{0}$00 cal

9. 92.0 cal

10. 101.2°C

11. (a) $\frac{1}{2}H_2(g) + \frac{1}{2}Cl_2(g) \rightarrow HCl(g) + 22.1$ kcal

(b) $\frac{1}{2}H_2(g) + \frac{1}{2}I_2(g) \rightarrow HI(g) - 6.2$ kcal

12. 24 kcal

13. 471 kcal

14. 14.3 kcal

15. $\Delta H_R = -472.6$

16. 251 kcal

17. ΔH_f° of CuO is −37.2 kcal/mole

18. 54,800 kcal

19. 52.5 kcal

20. $63\bar{0}0$ cal

21. d	22. a	23. c	24. a	25. c
26. d	27. d	28. b	29. c	30. d
31. b	32. b	33. d	34. b	35. a

CHAPTER 13 Quiz

1. How many liters of CO_2 are produced when 25.0 L of O_2 are used to burn propane? (Assume all gases are at the same temperature and pressure.)

   $C_3H_8 + O_2 \rightarrow CO_2 + H_2O$ (unbalanced)

2. Convert the following quantities:

   (a) 7600 torr = __________ atm (b) 0.500 atm = __________ torr

   (c) $-5\bar{0}$°C = __________ K (d) $1\bar{0}\bar{0}$ K = __________ °C

3. Find the volume occupied by 2.00 moles of nitrogen gas at STP.

4. Determine the molecular mass of a gas, given the following experimental data: Mass = 1.00 g, P = 1.50 atm, V = 82.0 mL, T = $3\bar{0}\bar{0}$ K.

5. A balloon is filled with a gas to a volume of $4\bar{0}\bar{0}$ mL at a pressure of 1.00 atm and a temperature of 27°C. What will the volume of the balloon be if the temperature is increased to 327°C and the pressure is decreased to $38\bar{0}$ torr?

6. Express the following pressures in atmospheres.
   (a) 155 torr (b) 4940 torr

7. Express the following pressures in torr.
   (a) 0.250 atm (b) 3.20 atm

8. The pressure of a gas is 3.00 atm and its volume is 8.00 liters. What is the volume of the gas if the pressure is increased to 12.00 atm? (Assume that the temperature remains constant.)

9. Change °C to K for each of the following temperatures:
   (a) 20°C (b) −40°C

10. Change K to °C for each of the following temperatures:
    (a) 6 K (b) 105 K

11. A gas has a volume of 800.0 mL at a temperature of 20.0°C. What is the volume of the gas if the temperature is increased to 100.0°C? (Assume that the pressure remains constant.)

12. A given amount of gas has a volume of 2.00 L at a pressure of 3.00 atm and a temperature of 100.0°C. What is the pressure of the gas if the temperature is increased to 473°C? (Assume that the volume of the gas is held constant.)

13. A gas has a volume of 9.80 L at a pressure of 1.50 atm and a temperature of 77.0°C. What is the volume of the gas at STP?

14. A sample of oxygen gas is collected over water. The volume of the gas collected is 400.0 mL at a pressure of 0.950 atm and a temperature of 15.0°C. What is the volume of the dry gas at STP? (P_{H_2O} at 15°C is 12.8 torr.)

15. Complete the following table:

Data set	P	V	n	T
1	Constant	Increased	Constant	? (a)
2	? (b)	Increased	Constant	Constant
3	Increased	? (c)	Decreased	Constant

16. Oxygen gas makes up about 20.0 percent by volume of the atmosphere. What is the pressure exerted by oxygen gas when the atmospheric pressure is $76\bar{0}$ torr?

17. Propane gas burns in air to produce carbon dioxide and water vapor.
(a) Write a balanced equation for this reaction.
(b) How many liters of oxygen gas are needed to burn 15.0 L of propane? (Assume all gases are at the same temperature and pressure.)
(c) How many liters of CO_2(g) and of water vapor are produced?

18. How many grams of neon gas are present in a 5.00-L container at a temperature of 127°C and a pressure of 1520.0 torr?

19. What volume will 0.550 mole of Kr gas occupy at STP?

20. Given that 4.00 g of a gas have a volume of 500.0 mL at a pressure of 2,005 torr and a temperature of 27.0°C, calculate the molecular mass of the gas.

21. If the temperature of a gas is increased at constant pressure its volume will:
(a) increase
(b) decrease
(c) remain the same
(d) Insufficient data to answer the question

22. A gas has a volume of 6.0 liters at a pressure of 380 torr. If the pressure is increased to 760 torr at the same temperature, what will be its new volume?
(a) 6.0 liters
(b) 3.0 liters
(c) 12 liters
(d) 9.0 liters

23. Which of the following is not a true statement about gases?
(a) They have no definite shape
(b) They have no definite volume
(c) They are not compressible
(d) Their volume decreases with increasing pressure

24. The volume of a gas varies directly with its kelvin temperature, at constant pressure. This is a statement of:
(a) Dalton's Law of Partial Pressures
(b) Gay Lussac's Law
(c) Boyle's Law
(d) Charles's Law

25. A gas has a volume of 2.00 liters at a temperature of 127°C. What will be the volume of the gas if the temperature is increased to 327°C? (Assume the pressure remains constant.)
(a) 4.00 liters
(b) 6.00 liters
(c) 2.00 liters
(d) 3.00 liters

26. If the temperature of a gas is decreased at constant pressure its volume will:
(a) increase
(b) decrease
(c) remain the same
(d) Insufficient data to answer the question

27. A gas has a volume of 9.0 liters at a pressure of 1520 torr. If the pressure is decreased to 380 torr, what will be its new volume? (Assume temperature remains constant.)
(a) 36 liters
(b) 0.28 liter
(c) 2.3 liters
(d) 4.6 liters

28. The fact that gas pressures are additive is known as:
(a) Dalton's Law of Partial Pressures
(b) Ideal Gas Law
(c) Boyle's Law
(d) Charles's Law

29. The fact that the volume of a gas varies inversely with pressure at constant temperature is known as:
(a) Dalton's Law of Partial Pressures
(b) Henry's Law
(c) Boyle's Law
(d) Charles's Law

30. The fact that the pressure of a gas varies inversely with volume at constant temperature is known as:
(a) Dalton's Law of Partial Pressures
(b) Henry's Law
(c) Boyle's Law
(d) Charles's Law

31. Which of the following is a statement of Boyle's Law?
(a) pressure and volume vary directly
(b) pressure and temperature vary directly
(c) pressure and temperature vary inversely
(d) pressure and volume vary inversely

32. Which of the following is a statement of Charles's Law?
    (a) pressure and volume vary directly
    (b) pressure and Celsius temperature vary directly
    (c) volume and kelvin temperature vary inversely
    (d) volume and kelvin temperature vary directly

33. A gas is heated from 27°C to 927°C at constant pressure. Its volume increases:
    (a) two-fold (b) three-fold
    (c) four-fold (d) five-fold

34. A gas has a volume of 10.0 L at 1520 torr pressure and a temperature of 27.0°C. How many moles of gas are present? (Note: PV = nRT and R = 0.0821 L-atm/°K-mole)
    (a) 0.812 mole (b) 81.2 moles
    (c) 617 moles (d) 6.17 moles

35. Determine the volume of 5.00 moles of O_2 gas at STP.
    (a) 4.48 L (b) 22.4 L (c) 44.8 L (d) 112 L

CHAPTER 13 Quiz Answers

1. 15.0 L
2. (a) $1\bar{0}$ atm (b) $38\bar{0}$ torr (c) 223 K (d) –173°C
3. 44.8 L
4. MM = $2\overline{00}$ g/mole
5. $16\bar{0}0$ mL
6. (a) 0.204 atm (b) 6.50 atm
7. (a) $19\bar{0}$ torr (b) 2430 torr
8. 2.00 L
9. (a) 293 K (b) 233 K
10. (a) –267°C (b) –160°C
11. 1020 mL
12. 6.00 atm
13. 11.5 L
14. 354 mL
15. (a) Increased (b) Decreased (c) Decreased
16. 152 torr

17. (a) $C_3H_8 + 5O_2 \rightarrow 3CO_2 + 4H_2O$
    (b) 75.0 L of O_2
    (c) 45.0 L of CO_2 and 60.0 L of H_2O

18. 6.10 g

19. 12.3 L

20. Molecular mass = 74.5

21. a	22. b	23. c	24. d	25. d
26. b	27. a	28. a	29. c	30. c
31. d	32. d	33. c	34. a	35. d

CHAPTER 14 Quiz

1. State whether each phrase that follows refers to a solid, a liquid, or a gas.
   (a) Has definite shape and definite volume.
   (b) Has no definite shape or volume.
   (c) Has no definite shape, but has definite volume.

2. It takes 15 minutes to cook a hard-boiled egg on a mountaintop. Will it take more than or less than 15 minutes to cook a hard-boiled egg at sea level?

3. Match each type of solid on the left with its characteristics on the right.
   (a) Ionic
   (b) Molecular
   (c) Metallic

   (1) High melting point, very hard, good conductor of electricity
   (2) High melting point, hard and brittle, does not conduct electricity
   (3) Low melting point, soft, does not conduct electricity

4. Define the following terms:
   (a) Normal boiling point
   (b) Equilibrium vapor pressure
   (c) Evaporation
   (d) Unit cell
   (e) Crystal lattice

5. Match each phrase on the left with the correct statement on the right.
   (a) Causes it to increase
   (b) Causes it to decrease
   (c) Has no effect on it

   (1) An increase in atmospheric pressure does this to the boiling point.
   (2) A decrease in atmospheric pressure does this to the melting point.
   (3) A decrease in atmospheric pressure does this to the boiling point.

6. True or False: A liquid has no definite shape or volume.

7. True or False: Gases are easily compressible, but solids and liquids are not.

8. True or False: A kettle of water is boiling on a stove. If you turn up the heat, the water gets hotter.

9. Solids that have no well-defined crystalline structure are said to be ________________.

10. Match each term on the left with its description on the right.

(a) Distillate	(1) The double-jacketed, water-cooled piece of glassware used in distillation
(b) Distilling head	(2) The condensed liquid
(c) Condenser	(3) The piece of glassware that connects the distilling flask to the double-jacketed water-cooled piece of glassware

11. Match each term on the left with its description on the right.

(a) Atoms	(1) In sodium chloride, these occupy the lattice positions.
(b) Molecules	(2) In graphite, these occupy the lattice positions.
(c) Ions	(3) In carbon dioxide solid (dry ice), these occupy the lattice positions.

12. True or False: Ionic solids are hard and brittle and have high melting points.

13. A certain compound is a soft solid with a low melting point. It does not conduct electricity. What type of particles occupy its lattice positions?

14. True or False: Molecular solids are nonconductors of electricity.

15. As water goes from a solid to a liquid to a gas, the energy of its molecules ________________ (increases or decreases).

16. When dry ice is placed in a beaker of water, vigorous bubbling occurs and a vapor leaves the beaker. Identify the vapor.

17. True or False: Glass is a crystalline solid.

18. True or False: Molecules of a liquid are constantly in motion.

19. True or False: The equilibrium vapor pressure of a liquid increases with temperature.

20. A substance that goes directly from a solid state to a gaseous state is said to ________________.

21. What compounds are obtained when water combines with each of the following?
(a) CO_2 (b) CaO (c) MgO (d) Na

22. Complete and balance the following equations involving reactions with water.
(a) $SO_3 + H_2O \rightarrow$ (b) $Li + H_2O \rightarrow$
(c) $Na_2O + H_2O \rightarrow$

23. Determine the percent by weight of water in $Na_2CO_3 \cdot 10H_2O$.

24. Match each term on the left with its definition on the right.

(a) Efflorescent	(1) A salt that has lost all of its water of hydration
(b) Deliquescent	(2) A substance that can absorb water from the atmosphere
(c) Anhydrous	(3) A substance that absorbs enough water from the atmosphere to form a solution
(d) Hygroscopic	(4) A salt that can spontaneously lose its water of hydration to the atmosphere

25. There are six forms of the hydrated salt $MnSO_4$. They range from $MnSO_4 \cdot 2H_2O$ to $MnSO_4 \cdot 7H_2O$. Given the following information, determine which of the hydrated salts is being analyzed: 20.00 g of hydrated salt weigh 12.53 g after heating.

26. Which of the following is not a true statement about liquids?
(a) the boiling point of a liquid will increase with increasing atmospheric pressure
(b) the boiling point of a liquid will decrease with decreasing atmospheric pressure
(c) liquids have no definite shape
(d) liquids have no definite volume

27. Which of the following is not a true statement about solids?
(a) the melting point of a solid is not affected by atmospheric pressure
(b) solids have definite shape
(c) solids have definite volume
(d) solids are easily compressible

28. Which of the following is a true statement about solids?
(a) the melting point of a solid is affected by atmospheric pressure
(b) solids have definite shape
(c) solids have no definite volume
(d) solids are easily compressible

29. The particles that occupy the lattice positions in a solid like diamond are:
(a) anions (b) cations
(c) atoms (d) molecules

30. An unknown solid has a low melting point. It is soft and does not conduct electricity. What type of solid is it?
(a) ionic (b) molecular
(c) atomic (d) metallic

31. The particles that occupy the lattice positions in a solid like table salt are:
(a) atoms (b) ions
(c) neutrons (d) molecules

32. An unknown solid has a high melting point. It is hard and brittle and does not conduct electricity. What type of solid is it?
(a) ionic (b) molecular
(c) metallic (d) nonmetallic

33. The particles that occupy the lattice positions in a solid like carbon dioxide (dry ice) are:
(a) atoms (b) molecules
(c) neutrons (d) ions

34. An unknown solid has a high melting point. It also conducts electricity. What type of solid is it?
(a) metallic
(b) molecular
(c) ionic
(d) nonmetallic

35. The particles that occupy the lattice positions in a solid like graphite are:
(a) atoms
(b) molecules
(c) neutrons
(d) ions

36. The boiling point of a liquid increases with:
(a) decreasing atmospheric pressure
(b) increasing atmospheric pressure
(c) increasing altitude
(d) does not change

37. A liquid that readily vaporizes has:
(a) many hydrogen bonds
(b) a high boiling point
(c) a high equilibrium vapor pressure
(d) a low equilibrium vapor pressure

38. Which of the following is an ionic solid:
(a) carbon dioxide
(b) cesium chloride
(c) diamond
(d) iodine

39. Where would it take the longest amount of time to cook a hard-boiled egg?
(a) New York City
(b) Denver
(c) Death Valley
(d) There is no difference between these three places.

40. Which of the following is a molecular solid?
(a) cesium chloride
(b) carbon dioxide
(c) graphite
(d) magnesium

41. In atomic solids, ______________ occupy the lattice positions.
(a) ions
(b) molecules
(c) atoms
(d) electrons

42. When water combines with MgO, the product formed is:
(a) $Mg(OH)_2$
(b) MgOH
(c) MgO_2
(d) MgH_2

43. The percentage of water by mass in the compound $CuSO_4 \cdot 5H_2O$ is:
(a) 13.9%
(b) 27.8%
(c) 18.1%
(d) 36.1%

44. A substance that absorbs water from the atmosphere, but remains a solid, is said to be
(a) efflorescent
(b) hygroscopic
(c) deliquescent
(d) anhydrous

45. A substance that absorbs enough water from the atmosphere to form a solution is said to be:
(a) efflorescent
(b) hygroscopic
(c) deliquescent
(d) anhydrous

CHAPTER 14 Quiz Answers

1. (a) Solid (b) Gas (c) Liquid
2. Less than 15 minutes. (The water will boil at a higher temperature at sea level.)
3. (1) c, (2) a, (3) b
4. See Chapter 14 of text or the Glossary for these definitions.
5. (1) a, (2) c, (3) b
6. False
7. True
8. False
9. amorphous
10. (1) c, (2) a, (3) b
11. (1) c, (2) a, (3) b
12. True
13. molecules
14. True
15. increases
16. CO_2
17. False
18. True
19. True
20. sublime
21. (a) H_2CO_3 (b) $Ca(OH)_2$ (c) $Mg(OH)_2$ (d) NaOH
22. (a) $SO_3 + H_2O \rightarrow H_2SO_4$
    (b) $2Li + 2H_2O \rightarrow 2LiOH + H_2$
    (c) $Na_2O + H_2O \rightarrow 2NaOH$
23. (a) 62.9 percent
24. (1) c, (2) d, (3) b, (4) a
25. $MnSO_4 \cdot 5H_2O$

26. d   27. d   28. b   29. c   30. b

31. b   32. b   33. b   34. a   35. a

36. b   37. c   38. b   39. b   40. b

41. c   42. a   43. d   44. b   45. c

CHAPTER 15 Quiz

1. True or False: "Like dissolves like."

2. Give an example of:
   (a) A solid-solid solution
   (b) A liquid-liquid solution
   (c) A gas-gas solution
   (d) A gas-liquid solution

3. Define: (a) Solute, (b) Solvent.

4. Given that 20.0 g of salt are dissolved in 60.0 g of water, find the concentration in percentage by mass.

5. What are the molarity and the normality of a solution made by dissolving 20.0 g of NaOH in enough water to make $50\overline{0}$ mL of solution?

6. Given the molarity of the following solutions of acids and bases, state their normality.
   (a) 2M H_2SO_4
   (b) 0.50M H_3PO_4
   (c) 5M $Ca(OH)_2$
   (d) 0.1M $Al(OH)_3$

7. You have 6.00N NaOH on the shelf. Explain how you would make $12\overline{0}$ mL of a 5.00N solution.

8. Determine the molality of a solution prepared by dissolving 18 g of glucose ($C_6H_{12}O_6$) in $50\overline{0}$ g of water.

9. Determine the freezing point of the solution in Question 8. The K_f for water is $1.86 \frac{^\circ\text{C kg-solvent}}{\text{mol solute}}$.

10. Determine the boiling point of a solution prepared by dissolving 248 g of ethylene glycol (MW = 62) in $25\overline{0}$ g of water. The K_b for water is $0.51 \frac{^\circ\text{C kg-solvent}}{\text{mol solute}}$.

11. True or False: Oxygen is the solvent in air.

12. You are preparing lemonade. You decide to add one more teaspoonful of sugar to the lemonade. However, this last teaspoonful of sugar does not dissolve. The lemonade solution is said to be ________________ .

13. What is the percent by mass of a solution prepared by dissolving 18.0 g of glucose in enough water to make 900.0 g of solution?

14. What is the molarity of the solution in Question 13? (Hint: Assume that the density of the solution is 1 g/mL. The MW of glucose is 180.)

15. What is the percent by volume of a solution prepared by dissolving 2.00 mL of ethanol in enough water to make 40.0 mL of solution?

16. How many grams of KCl are there in 200.0 mL of a 10.0 percent by mass-volume solution of KCl and water?

17. How many grams of KNO_3 are in 500.0 mL of a 2.00M solution?

18. What are the molarity and the normality of a solution prepared by dissolving 22.2 g of $Ca(OH)_2$ in enough water to prepare 800.0 mL of solution?

19. Determine the molality of a solution prepared by dissolving 17.1 g of $C_{12}H_{22}O_{11}$ in $5\overline{0}0$ g of water.

20. Determine the boiling point and the freezing point of the solution described in Question 19.

21. What is the percent mass-volume of a solution prepared by dissolving 9.00 g of alcohol in 30.0 mL of water?
(a) 3.00% (b) 30.0% (c) 0.333% (d) 33.3%

22. What is the molarity of a solution that contains 72 g of HCl in 8.0 liters of solution?
(a) 9.0M (b) 4.0M (c) 3.0M (d) 0.25M

23. Which of the following substances is a nonelectrolyte in aqueous solution?
(a) NaCl (b) $HC_2H_3O_2$
(c) $C_6H_{12}O_6$ (d) H_2SO_4

24. What is the percent mass-volume of a solution prepared by dissolving 25 g of sodium chloride in enough water to make 125 g of solution?
(a) $2\overline{0}$% (b) $5\overline{0}$% (c) 5.0% (d) 25%

25. What is the molarity of a solution prepared by dissolving 18.0 g of glucose ($C_6H_{12}O_6$) in $2\overline{0}\overline{0}$ mL of solution?
(a) 0.0900M (b) 2.00M
(c) 0.500M (d) 1.00M

26. How many grams of glucose are in 500.0 mL of a 8.00% by mass glucose in water solution?
(a) 8.00 g (b) 20.0 g (c) 30.0 g (d) 40.0 g

27. How many grams of NaOH are in 4.00 L of a 0.500M NaOH solution?
(a) 20.0 g (b) 80.0 g (c) 2.00 g (d) 8.00 g

28. What is the percent mass-volume of a solution prepared by dissolving 80.0 g of glucose in enough water to make 500.0 mL of solution?
(a) 0.160% (b) 10.0% (c) 16.0% (d) 1.00%

29. What is the molarity of a solution prepared by dissolving 18.0 g of $HC_2H_3O_2$ in enough water to make 600.0 mL of solution?
(a) 3.00M (b) 0.000500M
(c) 1.00M (d) 0.500M

30. How many grams of $Ca(OH)_2$ are in 800.0 mL of a 5.00M solution?
(a) 296 g (b) 4.00 g (c) 54.1 g (d) 0.400 g

31. A substance that ionizes completely in solution is a:
(a) weak electrolyte (b) strong electrolyte
(c) nonelectrolyte (d) weak acid

32. What is the molarity of an 5.00% mass/volume solution of ethyl alcohol (C_2H_6O) in water?
(a) 0.0500M (b) 5.00M
(c) 10.9M (d) 1.09M

33. How many mL of a 0.150M H_3PO_4 solution do you need to get 1.47 g of H_3PO_4?
(a) $1\overline{00}$ mL (b) 1.00 mL
(c) $1\overline{000}$ mL (d) 10.0 mL

34. What is the molarity of a 9.00% mass/volume solution of glucose ($C_6H_{12}O_6$) in water?
(a) 5.00M (b) 0.500M (c) 2.00M (d) 0.200M

35. How many mL of a 4.00M H_2SO_4 solution do you need to obtain 4.90 g of H_2SO_4?
(a) $1\overline{00}$ mL (b) 12.5 mL
(c) $1\overline{000}$ mL (d) 1250 mL

36. What is the molarity of a 4.00% mass/volume solution of sodium hydroxide (NaOH) solution in water?
(a) 1.00M (b) 2.00M (c) 3.00M (d) 4.00M

37. How many grams of $NaNO_3$ are needed to prepare 0.500 L of a 0.200M solution?
(a) 17.0 g (b) 85.0 g (c) 1.70 g (d) 8.50 g

38. What is the percent mass-volume 2.00M NaOH solution?
(a) 2.00% (b) 4.00% (c) 6.00% (d) 8.00%

39. How many grams of $Ca(NO_3)_2$ are needed to prepare 2.000 L of a 0.500M solution?
(a) 16.4 g (b) 164 g (c) 8.20 g (d) 82.0 g

40. What is the percent mass-volume 0.500M KOH solution?
(a) 2.80% (b) 0.500% (c) 28.0% (d) 5.00%

41. How many grams of $C_6H_{12}O_6$ are needed to prepare 500.0 mL of a 4.000M solution?
(a) 180.0 g (b) 360.0 g (c) 18.00 g (d) 36.00 g

42. The molality of a solution that contains 45.0 g of glucose, $C_6H_{12}O_6$, in 200 g of water is:
(a) 2.50 m (b) 5.00 m (c) 1.25 m (d) 0.225 m

43. The molality of a solution of ethylene glycol in water is 10.0 molal. The K_f of water is –1.86°C. What is the freezing point of the ethylene glycol/water solution?
(a) –10.0°C (b) 10.0°C
(c) –18.6°C (d) 18.6°C

44. A 3.00M H_2SO_4 solution is:
(a) 1.50N (b) 3.00N (c) 4.50N (d) 6.00N

45. The passage of a solvent through a semipermeable membrane is known as:
    (a) osmosis (b) diffusion
    (c) dialysis (d) dissolving

CHAPTER 15 Quiz Answers

1. True
2. There are many possible examples; see the text for some of them.
3. (a) Solute: the substance that is dissolved.
   (b) Solvent: the substance that does the dissolving of the solute.
4. 25.0 percent by mass
5. 1.00M and 1.00N
6. (a) 4N (b) 1.5N (c) 10N (d) 0.3N
7. Take 100 mL of the 6.00N NaOH and dilute to $12\overline{0}$ mL with water.
8. 0.20 molal
9. –0.37°C
10. 108.2°C
11. False
12. Saturated
13. 2.00 percent by mass
14. 0.111M
15. 5.00 percent by volume
16. 20.0 g of KCl
17. 101 g of KNO_3
18. 0.375M, 0.750N
19. 0.100 molal
20. The boiling point is 100.051°C. The freezing point is –0.186°C.
21. b 22. d 23. c 24. a 25. c
26. d 27. b 28. c 29. d 30. a
31. b 32. d 33. a 34. b 35. b

36. a 37. d 38. d 39. b 40. a

41. b 42. c 43. c 44. d 45. a

CHAPTER 16 Quiz

1. List three properties of acids, bases, and salts.

2. Complete and balance the following equations:
   (a) $Mg(OH)_2 + HNO_3 \rightarrow$ (b) $HCl + NH_4OH \rightarrow$
   (c) $Zn + HCl \rightarrow$ (d) $NaOH + SO_2 \rightarrow$
   (e) $Na_2O + H_2O \rightarrow$ (f) $SO_2 + H_2O \rightarrow$

3. Give the chemical name for each of the following:
   (a) Milk of magnesia (b) Ammonia water
   (c) Lye (d) Vinegar

4. Name the following acids, bases, and salts:
   (a) KOH (b) Na_2SO_4
   (c) H_3PO_4 (d) NaF
   (e) $Al(OH)_3$ (f) H_2CO_3

5. Write formulas for the following acids, bases, and salts:
   (a) Hydroiodic acid (b) Nitric acid
   (c) Calcium phosphate (d) Iron(II) hydroxide

6. Match each word on the left with its definition on the right.
   (a) Arrhenius acid (1) Proton donor
   (b) Arrhenius base (2) Releases OH^- ions in solution
   (c) Brønsted-Lowry acid (3) Releases H^+ ions in solution
   (d) Brønsted-Lowry base (4) Proton acceptor

7. Complete the following equations:
   (a) $NH_3 + H_2O \rightleftarrows ?$ (b) $(HSO_4)^{1-} \rightarrow ? + ?$

8. True or False: A base feels soapy or slippery to the touch.

9. Complete each sentence with the words "red" and "blue."
   (a) ________ lithmus paper turns ________ in acid.
   (b) ________ lithmus paper turns ________ in base.

10. Write an acid-base equation that shows the formation of potassium sulfate (K_2SO_4) as one of the products.

11. Write an equation showing the reaction of a metal oxide and water for the preparation of each of the following compounds:
    (a) Sodium hydroxide (b) Magnesium hydroxide

12. Write a balanced chemical equation for each of the following word equations:
    (a) Sulfuric acid + iron(III) hydroxide → iron(III) sulfate + water
    (b) Zinc + phosphoric acid → zinc phosphate + hydrogen gas

13. Write an equation for the ionization of water.

14. How many moles of hydrogen ions and hydroxide ions are present in 5.00 L of water?

15. How much more acidic than a substance whose pH = 6 is a substance whose pH = 2?

16. Determine the pH of the following aqueous solutions:
    (a) $[H^+] = 10^{-4}M$ (b) $[H^+] = 10^{-10}M$
    (c) $[OH^-] = 10^{-5}M$ (d) $[OH^-] = 10^{-9}M$

17. Determine the H^+ concentration of each of the following solutions:
    (a) pH = 3 (b) pH = 9
    (c) $[OH^-] = 10^{-2}M$ (d) $[OH^-] = 10^{-12}M$

18. Write a reaction showing the preparation of the salt zinc bromide via a single-replacement reaction.

19. Calculate the normality of a solution of HCl if it takes 30.00 mL of 0.500N NaOH solution to neutralize 40.00 mL of the HCl solution.

20. Calculate the percent by mass of acetic acid in vinegar if it takes 25.00 mL of 1.25M NaOH solution to neutralize a 40.00 g sample of vinegar.

21. A solution has an $[OH^-]$ of $10^{-9}M$. Its pH is:
    (a) 5 (b) 9 (c) 7 (d) 4

22. Which of the following is not a property of acids?
    (a) they release hydrogen ions in solution
    (b) they turn red litmus blue
    (c) they have a sour taste
    (d) they react with bases

23. A solution has an $[OH^-]$ of $10^{-6}M$. Its pH is:
    (a) 6 (b) 9 (c) 8 (d) 4

24. Which of the following is a property of acids?
    (a) they release hydrogen ions in solution
    (b) they turn red litmus blue
    (c) they have a bitter taste
    (d) they have pH's greater than 7

25. A substance that is composed of a metal plus a nonmetal is a(n):
    (a) acid (b) base (c) salt (d) oxide

26. A solution has an $[OH^-]$ of $10^{-12}M$. Its pH is:
    (a) 12 (b) 14 (c) 7 (d) 2

27. A 0.0001M solution of HCl has a pH of:
    (a) 3 (b) 4 (c) 11 (d) 10

28. A substance that is composed of a metal plus a hydroxide ion is a(n):
    (a) acid (b) base (c) salt (d) oxide

29. A substance that does not ionize in solution is a:
(a) weak electrolyte (b) strong electrolyte
(c) nonelectrolyte (d) weak base

30. A solution has an $[OH^-]$ of 10^{-5}M. Its pH is:
(a) 5 (b) 9 (c) 7 (d) 2

31. A 0.00001M solution of KOH has a pH of:
(a) 4 (b) 5 (c) 9 (d) 10

32. A substance that is composed of a hydrogen ion plus a nonmetal ion is a(n):
(a) acid (b) base (c) salt (d) oxide

33. A substance that partially ionizes in solution is a:
(a) nonelectrolyte (b) strong electrolyte
(c) weak electrolyte (d) weak base

34. A solution has an $[H^+]$ of 10^{-5}M. Its pH is:
(a) 5 (b) 9 (c) 7 (d) 2

35. A 0.0001M solution of NaOH has a pH of:
(a) 4 (b) 5 (c) 9 (d) 10

36. A solution has a pH of 3. Its hydrogen ion concentration is:
(a) 10^{-11}M (b) 10^{-7}M (c) 10^{-3}M (d) 10^{3}M

37. Substance A has a pH of 2, and substance B has a pH of 5. How much more acidic is substance A over substance B, with regard to hydrogen ion concentration?
(a) three times (b) ten times
(c) one hundred times (d) one thousand times

38. A solution has an $[OH^-]$ of 10^{-2}M. Its pH is:
(a) 2 (b) 12 (c) 10 (d) 7

39. A 0.000001M solution of KOH has a pH of:
(a) 6 (b) 3 (c) 8 (d) 10

40. A solution has a pH of 8. Its hydroxide ion concentration is:
(a) 10^{-6}M (b) 10^{-8}M (c) 10^{-3}M (d) 10^{3}M

41. Substance A has a pH of 3, and substance B has a pH of 7. How much more acidic is substance A over substance B, with regard to hydrogen ion concentration?
(a) four times (b) one hundred times
(c) one thousand times (d) ten thousand times

42. A solution has an $[OH^-]$ of 10^{-1}M. Its pH is:
(a) 2 (b) 12 (c) 13 (d) 6

43. A 0.001M solution of CsOH has a pH of:
(a) 6 (b) 3 (c) 8 (d) 11

44. A solution has a pH of 3. Its hydroxide ion concentration is:
(a) 10^{-6}M (b) 10^{-8}M (c) 10^{-3}M (d) 10^{-11}M

45. Substance A has a pH of 8, and substance B has a pH of 12. How much more alkaline is substance B over substance A, with regard to hydroxide ion concentration?
(a) four times (b) one hundred times
(c) one thousand times (d) ten thousand times

46. A solution has an $[OH^-]$ of $10^{-1}M$. Its pH is:
(a) 1 (b) 12 (c) 13 (d) 7

47. A 0.1M solution of KOH has a pH of:
(a) 13 (b) 12 (c) 1 (d) 2

CHAPTER 16 Quiz Answers

1. There are many possible answers. See the text.

2. (a) $Mg(OH)_2 + 2HNO_3 \rightarrow Mg(NO_3)_2 + 2H_2O$
(b) $HCl + NH_4OH \rightarrow NH_4Cl + H_2O$
(c) $Zn + 2HCl \rightarrow ZnCl_2 + H_2$
(d) $2NaOH + SO_2 \rightarrow Na_2SO_3 + H_2O$
(e) $Na_2O + H_2O \rightarrow 2NaOH$
(f) $SO_2 + H_2O \rightarrow H_2SO_3$

3. (a) Magnesium hydroxide (b) Ammonium hydroxide
(c) Sodium hydroxide (d) Acetic acid

4. (a) Potassium hydroxide (b) Sodium sulfate
(c) Phosphoric acid (d) Sodium fluoride
(e) Aluminum hydroxide (f) Carbonic acid

5. (a) HI (b) HNO_3 (c) $Ca_3(PO_4)_2$ (d) $Fe(OH)_2$

6. (1) c, (2) b, (3) a, (4) d

7. (a) $NH_3 + H_2O \rightleftarrows (NH_4)^{1+} + (OH)^{1-}$
(b) $(HSO_4)^{1-} \rightarrow H^{1+} + (SO_4)^{2-}$

8. True

9. (a) Blue, red (b) Red, blue

10. $2KOH + H_2SO_4 \rightarrow K_2SO_4 + 2H_2O$

11. (a) $Na_2O + H_2O \rightarrow 2NaOH$ (b) $MgO + H_2O \rightarrow Mg(OH)_2$

12. (a) $3H_2SO_4 + 2Fe(OH)_3 \rightarrow Fe_2(SO_4)_3 + 6H_2O$
(b) $3Zn + 2H_3PO_4 \rightarrow Zn_3(PO_4)_2 + 3H_2$

13. $H_2O + H\text{—}OH \rightleftarrows H_3O^{1+} + OH^{1-}$

14. In 5.00 L of water, there are 5×10^{-7} mole of OH^{1-} ions and 5×10^{-7} mole of H^{1+} ions.

15. It is 10^4, or 10,000, times more acidic.

16. (a) pH = 4 (b) pH = 10 (c) pH = 9 (d) pH = 5

17. (a) $[H^+] = 10^{-3}M$ (b) $[H^+] = 10^{-9}M$
(c) $[H^+] = 10^{-12}M$ (d) $[H^+] = 10^{-2}M$

18. $Zn + 2HBr \rightarrow ZnBr_2 + H_2$

19. 0.375N

20. 4.69 percent by mass

21. a	22. b	23. c	24. a	25. c
26. d	27. b	28. b	29. c	30. b
31. c	32. a	33. c	34. a	35. d
36. c	37. d	38. b	39. c	40. a
41. d	42. c	43. d	44. d	45. d
46. c	47. a			

CHAPTER 17 Quiz

1. Match each term on the left with its definition on the right.

(a) Chemical kinetics
(b) Chemical equilibrium
(c) LeChatelier's principle
(d) Law of Mass Action

(1) A system under stress that will move in a direction to relieve that stress.
(2) The study of the rate of speed at which a chemical reaction occurs
(3) The rate of a chemical reaction is proportional to the concentration of the reacting species.
(4) A dynamic state in which two or more opposing processes are taking place simultaneously at the same rate

2. True or False:
(a) In a reaction at equilibrium, the concentrations of the reactants and products are the same.

(b) A catalyst increases reaction rates by changing reaction mechanisms.

(c) The amount of product obtained at equilibrium is proportional to how fast equilibrium is attained.

(d) A catalyst lowers the activation energy of a reaction by equal amounts for both forward and reverse reactions.

3. Calculate K_{eq} at 440°C for the reaction

   $H_2(g) + I_2(g) = 2HI(g)$

   The concentrations at equilibrium are as follows: $H_2(g)$ = 0.218 mole/liter, $I_2(g)$ = 0.218 mole/liter, and HI(g) = 1.564 mole/liter.

4. Calculate the equilibrium concentration of H^{1+} and $C_2H_3O_2^{1-}$ at 25°C for a 0.100M acetic acid solution. ($K_a = 1.80 \times 10^{-5}$)

5. Calculate the solubility of Ag_2CrO_4 at 25°C in grams per liter, given that K_{sp} = 1.9×10^{-12}.

6. True or False: At equilibrium, the rate of the forward reaction equals the rate of the reverse reaction.

7. True or False: Adding a catalyst to a reaction at equilibrium shifts the reaction to the right (that is, there will be more product).

8. Name four factors that affect the rate of a chemical reaction.

9. A reaction occurs in $2\bar{0}$ seconds at 10.0°C. How quickly will the reaction occur at 30.0°C?

10. Write the equilibrium equation for acetic acid in water.

11. Write the equilibrium expression for each of the following equations.

    (a) $3A(g) + 4B(g) \rightleftarrows 5C(g) + 6D(g)$

    (b) $2NO_2(g) \rightleftarrows N_2O_4(g)$

12. Calculate K_b for NH_3 at 20°C for the reaction:

    $NH_3(aq) + H_2O(aq) \rightleftarrows (NH_4)^{1+}_{(aq)} + (OH)^{1-}_{(aq)}$

    The equilibrium concentrations are as follows:

    $[(NH_4)^{1+}]$ = 0.0026M, $[(OH)^{1-}]$ = 0.0026M, and $[NH_3]$ = 0.188M.

13. Calculate the concentration of H^{1+} and F^{1-} in a 0.600M HF solution. ($K_a = 8.29 \times 10^{-4}$)

14. The pH of an acetic acid solution is 2.0. Write the equilibrium expression for acetic acid, and determine the concentration of acetic acid that gave this pH. ($K_a = 1.8 \times 10^{-5}$)

15. Calculate the equilibrium concentrations of $(CuOH)^{1+}$ and $(OH)^{1-}$ for a 0.0100M copper(II) hydroxide solution. ($K_b = 1.00 \times 10^{-8}$)

16. What is the pH of the solution described in Question 15?

17. Write K_{sp} expressions for the following slightly soluble salts:
    (a) $Al(OH)_3$ (b) PbI_2

18. Calculate the K_{sp} for calcium fluoride if the concentration of Ca^{2+} is 1.07×10^{-4}M and the concentration of F^{1-} is 6.05×10^{-4}M.

19. Calculate the solubility of $Fe(OH)_3$ in grams per liter at 25°C. The K_{sp} of $Fe(OH)_3$ is 6.0×10^{-38}.

20. Consider the equation

$H_2O + CO_2 \rightleftarrows H_2CO_3 \rightleftarrows H^+ + HCO_3^-$

Decide which way the equilibrium shifts when:
(a) CO_2 is removed.
(b) Excess H^{1+} is added.
(c) HCO_3^{1-} is removed.

21. A system under stress will move in a direction to relieve that stress is an example of:
(a) chemical kinetics
(b) chemical equilibrium
(c) LeChatelier's principle
(d) Law of Mass Action

22. The study of the rate at which a chemical reaction occurs is the study of:
(a) chemical kinetics
(b) chemical equilibrium
(c) LeChatelier's principle
(d) Law of Mass Action

23. A dynamic state in which two or more opposing processes are taking place simultaneously at the same rate is:
(a) chemical kinetics
(b) chemical equilibrium
(c) LeChatelier's principle
(d) Law of Mass Action

24. A reaction occurs in 20 seconds at a temperature of 20°C. Approximately how long will it take the same reaction to occur at a temperature of 40°C?
(a) 5 seconds
(b) 10 seconds
(c) 40 seconds
(d) 80 seconds

25. A catalyst affects reaction rates by:
(a) adding energy to the reactants
(b) adding energy to the products
(c) lowering the activation energy
(d) shifting the equilibrium of the reaction to the product side of the equation

26. The equilibrium expression for the reaction

$2A(g) + 3B(g) = 5C(g) + 4D(g)$

is:

(a) $K_{eq} = [B]^3[A]^2/[C]^5[D]^4$

(b) $K_{eq} = [D]^4[A]^2/[B]^3[C]^5$

(c) $K_{eq} = [D]^4[B]^3/[A]^2[C]^5$

(d) $K_{eq} = [D]^4[C]^5/[B]^3[A]^2$

27. Given the following reaction:

$A(g) + 2B(g) = 2C(g)$

and the concentration of A(g) = 2.00M, B(g) = 2.00M, and C = 6.00M, what is the K_{eq} for the reaction?
(a) 9.80
(b) 6.00
(c) 19.6
(d) 4.50

28. The K_a for acetic acid ($HC_2H_3O_2$) is 1.80×10^{-5} at 20°C. The hydrogen ion concentration for a 0.100M solution of this acid is:
(a) 1.80×10^{-6}M
(b) 1.34×10^{-3}M
(c) 5.50×10^{-5}M
(d) 7.46×10^{-3}M

29. The K_{sp} expression for $Fe(OH)_3$ is:
(a) $K_{sp} = [Fe^{3+}][OH^{1-}]$
(b) $K_{sp} = [Fe^{3+}][3OH^{1-}]$
(c) $K_{sp} = [Fe^{3+}][OH^{1-}]^3$
(d) $K_{sp} = [Fe^{3+}]^3[OH^{1-}]$

30. The solubility product constant for Ag_2CrO_4 at 20°C is 1.9×10^{-12}. Calculate the solubility of Ag_2CrO_4 in grams per liter.
(a) 2.6×10^{-2} g/L
(b) 1.3×10^{-2} g/L
(c) 1.1×10^{-2} g/L
(d) 6.8×10^{-2} g/L

31. For the following reaction at equilibrium:

$H_2(g) + Cl_2(g) = 2HCl(g) + 44.2$ kcal

which way does the equilibrium shift, if at all, if heat is added to the system?
(a) left
(b) right
(c) no shift
(d) Insufficient information to answer this question

32. For the following reaction at equilibrium:

$H_2(g) + Cl_2(g) = 2HCl(g) + 44.2$ kcal

which way does the equilibrium shift, if at all, if pressure is increased on the system?
(a) left
(b) right
(c) no shift
(d) Insufficient information to answer this question

33. Which of the following equilibrium constants indicates that the reaction goes furthest to completion?
(a) $K_{eq} = 1$
(b) $K_{eq} = 10$
(c) $K_{eq} = 100$
(d) $K_{eq} = 1000$

34. An effect of hyperventilation is:
(a) the pH of blood increases
(b) the pH of blood decreases
(c) the concentration of H^+ decreases
(d) respiratory acidosis

35. Which of the following does not affect the equilibrium of a reaction?
(a) concentration of reactants
(b) temperature
(c) catalyst
(d) concentration of products

CHAPTER 17 Quiz Answers

1. (1) c, (2) a, (3) d, (4) b

2. (a) False (b) True (c) False (d) True

3. $K_{eq} = \frac{[HI]^2}{[H_2][I_2]} = \frac{(1.564)^2}{(0.218)^2} = 51.4$

4. $K_a = \frac{[H^{1+}][Ac^{1-}]}{[HAc]}$

   $1.8 \times 10^{-5} = \frac{x^2}{0.100}$ $x^2 = 1.8 \times 10^{-6}$

   $x = [H^{1+}] = [Ac^{1-}] = 1.34 \times 10^{-3}$

5. $Ag_2CrO_4 = 2Ag^{1+} + CrO_4^{2-}$

   $K_{sp} = [Ag^{1+}]^2[CrO_4^{2-}]$ $1.9 \times 10^{-12} = (2x)^2(x) = 4x^3$

   $x = 7.8 \times 10^{-5}$ mole/L (which also equals the amount of dissolved Ag_2CrO_4)

   ? g/L of dissolved Ag_2CrO_4

   $= (7.8 \times 10^{-5}$ mole/L)(332 g/mole) $= 2.6 \times 10^{-2}$ g/L

6. True

7. False

8. Nature of the reactants, concentration, temperature, the presence of catalysts, particle size, stirring, etc.

9. 5 seconds

10. $HC_2H_3O_2 + H_2O \rightleftarrows H_3O^{1+} + C_2H_3O_2^{1-}$

11. (a) $K_{eq} = \frac{[D]^6[C]^5}{[A]^3[B]^4}$ (b) $K_{eq} = \frac{[N_2O_4]}{[NO_2]^2}$

12. 3.6×10^{-5}

13. 2.23×10^{-2}M

14. $K_{eq} = \frac{[H^{1+}][C_2H_3O_2^{1-}]}{[HC_2H_3O_2]}$, $[HC_2H_3O_2] = 5.6$M

15. $[(CuOH)^{1+}] = [(OH)^{1-}] = 1.00 \times 10^{-5}$M

16. pH = 9.0

17. (a) $K_{sp} = [Al^{3+}][OH^{1-}]^3$ (b) $K_{sp} = [Pb^{2+}][I^{1-}]^2$

18. $K_{sp} = 3.92 \times 10^{-11}$

19. 2.3×10^{-8} g/L

20. (a) The equilibrium shifts left.
    (b) The equilibrium shifts left.
    (c) The equilibrium shifts right.

21. c	22. a	23. b	24. a	25. c
26. d	27. d	28. b	29. c	30. a
31. a	32. c	33. d	34. a	35. c

CHAPTER 18 Quiz

1. List alpha, beta, and gamma radiation in order from least penetrating to most penetrating.

2. Complete the following nuclear equations:
   (a) ${}^{212}_{84}Po \rightarrow {}^{4}_{2}He +$ ________ (b) ${}^{238}_{92}U + {}^{1}_{0}n \rightarrow {}^{94}_{36}Kr +$ ________

3. The half-life of a particular isotope is 2 hours. You have $\overline{1000}$ g of this isotope today at 12:00 noon. How many grams will be left by 6:00 P.M. tonight?

4. Define the following terms:
   (a) Voltaic cell (b) Electrolytic cell
   (c) Fuel cell

5. What type of pollution-control device would you place on a smokestack:
   (a) Of a city incinerator that is spewing out black smoke?
   (b) Of a power company that is using high-sulfur fuel for generating electricity?

6. Match each term on the left with its definition on the right:

(a) Alpha	(1) The type of radiation that resembles an electron
(b) Beta	(2) The type of radiation that is the most penetrating
(c) Gamma	(3) The type of radiation that resembles a helium atom with its electrodes removed

7. Complete the following nuclear equations:
   (a) ${}^{234}_{90}Th \rightarrow {}^{0}_{-1}e + ?$ (b) ${}^{214}_{83}Bi \rightarrow {}^{4}_{2}He + ?$

8. Complete the following sentence with "increases," "decreases," or "does not change."

   When an isotope emits an alpha particle, its mass number ________ and its atomic number ________.

9. What isotope forms when ${}^{194}_{78}Pt$ emits an alpha particle from its nucleus?

10. The half-life of cobalt-60 is 5.3 years. If you had a pure sample of cobalt-60 today that weighed $1\overline{000}$ g, how much would be left after 21.2 years?

11. The isotope iodine-131 has a half-life of 8 days. Suppose a sample of iodine-131 originally weighed $1\overline{0}\overline{0}$ g. How old is the sample today if the amount of iodine-131 remaining weighs only 6.25 g?

12. Complete the following sentence with the words "fission" and "fusion."

    Splitting the atom is nuclear ______________, whereas combining hydrogen nuclei to form helium nuclei is nuclear ______________.

13. Match each term on the left with its description on the right.

(a) Teletherapy	(1) A radioactive isotope is placed in the area to be treated.
(b) Brachytherapy	(2) A high-energy beam of radiation is aimed at cancerous tissue.
(c) Radiopharmaceutical therapy	(3) A radioactive isotope is administered either orally or intravenously.

14. Balance the following equation for a discharging lead storage battery.

    $Pb + PbO_2 + H_2SO_4 \rightarrow PbSO_4 + H_2O$

15. Match each term on the left with its definition on the right.

(a) Voltaic cell	(1) A cell in which reactants are introduced continually and products are removed continually
(b) Electrolytic cell	(2) A cell in which electricity is used to produce a chemical reaction
(c) Fuel cell	(3) An ordinary flashlight battery

16. Complete and balance the following equations, which are involved in the formation of acid rain:

    (a) $SO_2 + O_2 \rightarrow$ (b) $H_2O + SO_3 \rightarrow$

17. Match each term on the left with the example of it on the right.

(a) Particulate pollutant	(1) PCBs
(b) Gaseous pollutant	(2) Fly ash
(c) Hazardous waste	(3) Sulfur oxides

18. What type of control device is used to stop the emission of sulfur oxide?

19. True or False: A CAT scan uses computers to analyze radioactive isotopes that have been absorbed in body tissue.

20. True or False: The unit known as the curie expresses the activity of a radiation source.

21. An alpha particle can be represented as:

    (a) $^{0}_{-1}e$ (b) $^{4}_{2}He$ (c) $^{0}_{1}n$ (d) $^{1}_{1}p$

22. For the following equation:

    $^{14}_{6}C \rightarrow ^{0}_{-1}e + ?$

    The missing isotope is:

    (a) $^{14}_{5}B$ (b) $^{13}_{6}C$ (c) $^{15}_{6}C$ (d) $^{14}_{7}N$

23. The isotope ^{90}Sr has a half-life of 28 years. If you have 16 g of ^{90}Sr today, how much will be left in 112 years?
(a) 8.0 g (b) 4.0 g (c) 2.0 g (d) 1.0 g

24. A beta particle can be represented as:
(a) $^{0}_{-1}e$ (b) $^{4}_{2}He$ (c) $^{0}_{1}n$ (d) $^{1}_{1}p$

25. For the following equation:

$^{87}_{36}Kr \rightarrow ^{1}_{0}n + ?$

the missing atom is:
(a) $^{85}_{36}Rb$ (b) $^{85}_{37}Rb$ (c) $^{86}_{36}Kr$ (d) $^{88}_{37}Kr$

26. The isotope I-131 has a half-life of eight days. If you have 40.0 g of I-131 today, how many grams will you have left in 24 days?
(a) 5.00 g (b) 10.0 g (c) 15.0 g (d) 20.0 g

27. An excellent device used for removing particulate matter in a smokestack is:
(a) a scrubber (b) an electrostatic precipitator
(c) a water spray (d) none of these

28. When a radioactive atom emits an alpha particle, its mass number:
(a) increases by two (b) decreases by two
(c) increases by four (d) decreases by four

29. For the following equation:

$^{226}_{88}Ra \rightarrow ^{222}_{86}Ra + ?$

the missing particle is:
(a) $^{0}_{-1}e$ (b) $^{4}_{2}He$ (c) $^{1}_{1}p$ (d) $^{1}_{0}n$

30. The isotope ^{14}C has a half-life of 5770 years. A piece of wood originally contained 4.00 g of ^{14}C. Today it contains only 1.00 g of ^{14}C. Abut how old is the piece of wood?
(a) 5770 years (b) 11,540 years
(c) 17,310 years (d) 23,080 years

31. Which type of nuclear radiation has no charge associated with it?
(a) alpha (b) beta (c) gamma (d) proton

32. The emission of gamma radiation causes the emitting isotope to:
(a) increase its atomic number by one
(b) decrease its mass number by four
(c) lose energy
(d) decrease its mass number by two

33. For the following equation:

$^{3}_{1}H + ^{2}_{1}H \rightarrow ^{4}_{2}He + ?$

the missing particle is:

(a) $^{1}_{0}n$ (b) $^{1}_{+1}p$ (c) $^{0}_{-1}e$ (d) $^{4}_{+2}He$

34. The isotope Sr-90 has a half-life of 28 years. If you have 2000.0 g of Sr-90 today, how many grams will you have left in 112 years?
(a) 500.0 g (b) 250.0 g (c) 125.0 g (d) 62.50 g

35. Acid rain typically has a pH of:
(a) exactly 7.0
(b) greater than 8.0
(c) less than 6.0
(d) exactly zero

36. For the nuclear reaction:

$^{239}_{94}Pu \rightarrow ^{235}_{92}U + ?$

the missing particle is:

(a) $^{0}_{-1}e$ (b) $^{1}_{0}n$ (c) $^{4}_{+2}He$ (d) $^{1}_{+1}p$

37. The following nuclear equation:

$^{2}_{1}H + ^{3}_{1}H \rightarrow ^{4}_{2}He + ^{1}_{0}n$

is an example of:

(a) a chain reaction
(b) nuclear fission
(c) nuclear fusion
(d) a branching chain

38. Molecules of water attract each other due to:
(a) covalent bonds
(b) hydrogen bonds
(c) ionic bonds
(d) ion pairs

39. A sample of air is analyzed and found to contain 0.000200 g of particulate matter in 40.0 m^3 of air. What is the concentration of particulate matter in micrograms/cubic meter ($\mu g/m^3$)?
(Hint: 1 $\mu g = 1 \times 10^{-6}$ g)

(a) 5.00 $\mu g/m^3$
(b) 0.200 $\mu g/m^3$
(c) 50.0 $\mu g/m^3$
(d) 2.00 $\mu g/m^3$

40. Which of the following air pollution devices controls gaseous air pollutants from a smokestack?
(a) baghouse
(b) electrostatic precipitator
(c) scrubber
(d) cyclone

41. For the nuclear reaction:

$^{234}_{90}Th \rightarrow ^{234}_{91}Pa + ?$

the missing particle is:

(a) $^{0}_{-1}e$ (b) $^{1}_{0}n$ (c) $^{4}_{+2}He$ (d) $^{1}_{+1}p$

42. A unit of radiation that describes the activity of a radioactive source is the:
(a) rad (b) rem (c) röentgen (d) curie

43. A sample of air is analyzed for lead, and the concentration is found to be 2.50 $\mu g/m^3$. How many grams of lead would be present in 100.0 m^3 of air? (Hint: 1 $\mu g = 1 \times 10^{-6}$ g)
    (a) 2.50×10^{-4} g
    (b) 2.50×10^{-3} g
    (c) 2.50×10^{-2} g
    (d) 2.50×10^{-1} g

44. The major component of our atmosphere is:
    (a) oxygen
    (b) nitrogen
    (c) hydrogen
    (d) carbon dioxide

45. For the nuclear reaction:

    $^{2}_{1}H + ^{3}_{1}H \rightarrow ^{4}_{2}He + ?$

    the missing particle is:
    (a) $^{0}_{-1}e$
    (b) $^{1}_{0}n$
    (c) $^{4}_{+2}He$
    (d) $^{1}_{+1}p$

46. A unit of radiation that measures the ionizing ability of X rays and gamma rays is the:
    (a) rad
    (b) rem
    (c) röentgen
    (d) curie

47. A sample of air is analyzed for ozone, and the concentration is found to be 5.00 $\mu g/m^3$. How many grams of ozone would be present in 100.0 m^3 of air? (Hint: 1 $\mu g = 1 \times 10^{-6}$ g)
    (a) 5.00×10^{-4} g
    (b) 5.00×10^{-3} g
    (c) 5.00×10^{-2} g
    (d) 5.00×10^{-1} g

48. For the nuclear reaction:

    $^{6}_{3}Li + ^{1}_{0}n \rightarrow ^{3}_{1}H + ?$

    the missing particle is:
    (a) $^{0}_{-1}e$
    (b) $^{1}_{0}n$
    (c) $^{4}_{+2}He$
    (d) $^{1}_{+1}p$

49. The unit of radiation that has a weighting factor is the
    (a) rad
    (b) rem
    (c) röentgen
    (d) curie

50. The hydrogen bomb uses the principle of:
    (a) nuclear fusion
    (b) nuclear fission
    (c) low temperature fusion
    (d) low temperature fission

51. A fresh sample of ^{99}Tc is delivered to a laboratory. The half-life of ^{99}Tc is 6 hours. How long will it take for the activity of the sample to be reduced to one-fourth (¼) of its original activity?
    (a) 6 hours
    (b) 12 hours
    (c) 18 hours
    (d) 24 hours

52. An ordinary flashlight battery is an example of a:
    (a) fuel cell
    (b) voltaic cell
    (c) electrolytic cell
    (d) none of the above

CHAPTER 18 Quiz Answers

1. Alpha is the least penetrating, then beta, and gamma is the most penetrating.

2. (a) $^{212}_{84}Po \rightarrow ^{4}_{2}He + ^{208}_{82}Pb$ (b) $^{238}_{92}U + ^{1}_{0}n \rightarrow ^{94}_{36}Kr + ^{145}_{56}Ba$

3. 125 g

4. See Chapter 18 of the text or the Glossary for definitions.

5. (a) An electrostatic precipitator
   (b) An electrostatic precipitator and a scrubber

6. (1) b, (2) c, (3) a

7. (a) $^{234}_{91}Pa$ (b) $^{210}_{81}Tl$

8. decreases, decreases

9. $^{190}_{76}Os$

10. 62.5 g

11. 32 days old

12. fission, fusion

13. (1) b, (2) a, (3) c

14. $Pb + PbO_2 + 2H_2SO_4 \rightarrow 2PbSO_4 + 2H_2O$

15. (1) c, (2) b, (3) a

16. (a) $2SO_2 + O_2 \rightarrow 2SO_3$ (b) $H_2O + SO_3 \rightarrow H_2SO_4$

17. (1) c, (2) a, (3) b

18. A scrubber

19. False

20. True

21. b 22. d 23. d 24. a 25. c

26. a 27. b 28. d 29. b 30. b

31. c 32. c 33. a 34. c 35. c

36. c 37. c 38. b 39. a 40. c

41. a 42. d 43. a 44. b 45. b

46. c 47. a 48. c 49. b 50. a

51. b 52. b

CHAPTER 19 Quiz

1. Name the two crystalline forms of carbon.

2. Write the three isomers of pentane.

3. Write the general formulas for:
   (a) Alkanes (b) Alkenes
   (c) Alkynes

4. Name the following compounds, using the IUPAC rules:
   (a) 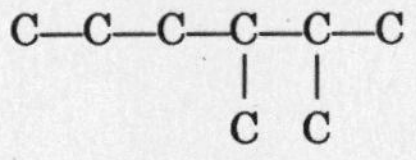
   (b) 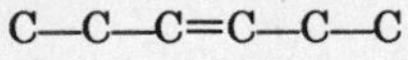
   (c) 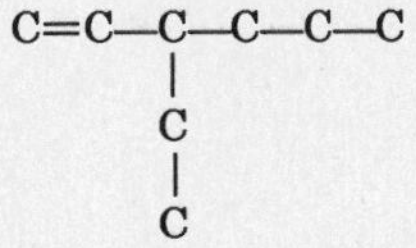
   (d) 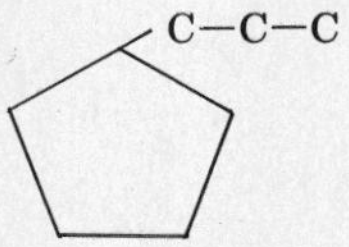

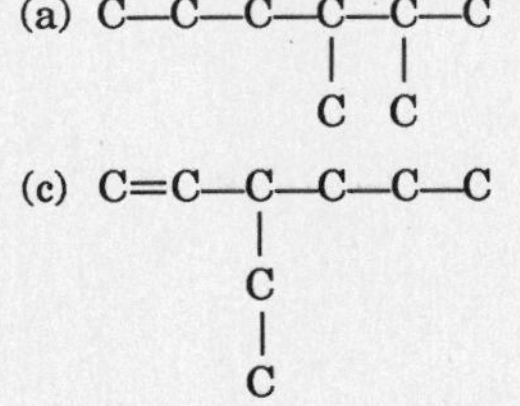

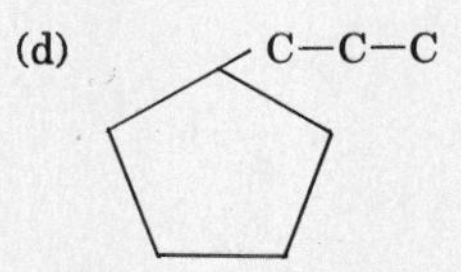

5. Write the formula for each of the following compounds:
   (a) 3-methylheptane (b) 2-methyl-4-ethylhexane
   (c) 3-methyl-1-ethylcyclohexane (d) 3-methyl-1-butene

6. True or False: Carbon-carbon bonds in organic compounds are ionic.

7. True or False: The binding pattern of the carbon atoms in many organic compounds is tetrahedral.

8. Draw the isomers of C_4H_9Cl.

9. Match each term on the left with its description on the right.
   (a) Alkane (1) Contains a carbon-carbon double bond
   (b) Alkene (2) Contains carbon-carbon single bonds
   (c) Alkyne (3) Contains a carbon-carbon triple bond

10. Rewrite the following hydrocarbon by adding hydrogen atoms to the carbon skeleton.

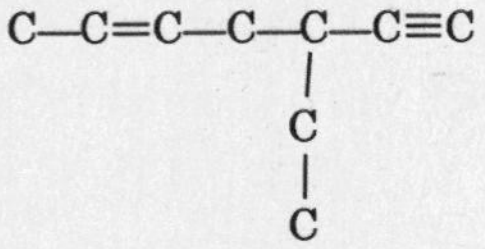

11. Write the structural formula for each of the following hydrocarbons:
    (a) 2-methylpropane (b) 3-ethyl-2-methylhexane

12. Name the first five straight-chain members of the alkene series.

13. Name each of the following alkenes:

(a)
```
C—C—C=C—C
```

(b)
```
C—C—C=C—C—C—C
  |       |
  C       C
```

14. Name each of the following alkynes:

(a)
```
            C
            |
C—C≡C—C—C—C
         |  |
         C  C
```

(b)
```
C—C—C≡C—C—C—C—C
  |         |
  C         C
```

15. Name each of the following cyclic hydrocarbons:

(a)

(b) CH_3

(c)

(d) CH_3 CH_3

16. Write the structural formula for each of the following alkanes:
(a) 2,2,3-trimethylhexane (b) 3-ethyl-3-methylpentane

17. Write the structural formula for each of the following alkenes:
(a) 3-ethyl-2-pentene (b) 4-ethyl-4-methyl-1-hexene

18. Write the structural formula for each of the following alkynes:
(a) 3,5,5-trimethyl-1-nonyne (b) 3-ethyl-1-pentyne

19. Write the structural formula for each of the following cyclic hydrocarbons:
(a) 1,2-dimethylcyclopentane (b) 4-methylcyclohexene

20. Write the structural formula for each of the following aromatic hydrocarbons:
(a) Benzene (b) Toluene

21. In organic compounds, carbon usually forms:
(a) one bond
(b) two bonds
(c) three bonds
(d) four bonds

22. Which of the following hydrocarbons is not a structural isomer of the others?

(a) 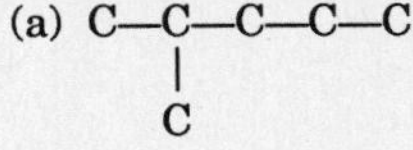

(b) 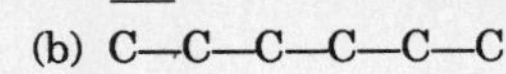

(c) 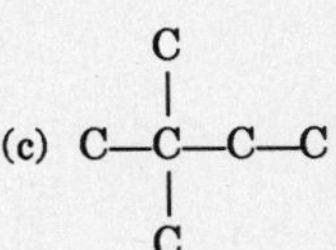

(d) 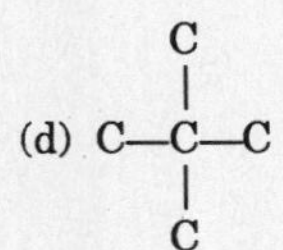

23. In a carbon-carbon triple bond, the two tetrahedral carbon atoms share a common:

(a) point
(b) edge
(c) side
(d) none of these

24. Which of the following compounds is not a structural isomer of the others?

(a) 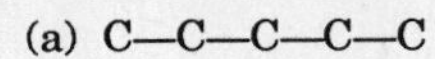

(b) 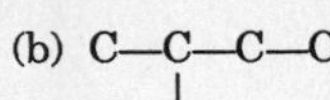

(c) 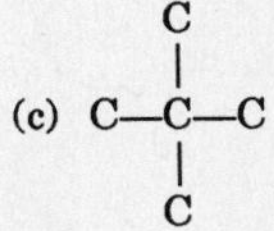

(d) 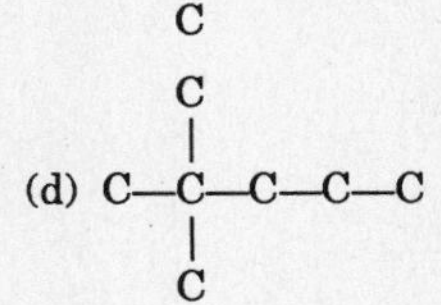

25. The general formula for an alkane is:

(a) C_nH_{2n} (b) C_nH_{2n+2} (c) C_nH_{2n-2} (d) C_nH_{3n-2}

26. The general formula for an alkyne is:

(a) C_nH_{2n} (b) C_nH_{2n+2} (c) C_nH_{2n-2} (d) C_nH_{3n-2}

27. The compound C—C—C≡C—C, whose carbon skeleton is shown, is a member of the:

(a) alkane series
(b) alkene series
(c) alkyne series
(d) cycloalkane series

28. The general formula for an alkene is:

(a) C_nH_{2n} (b) C_nH_{2n+2} (c) C_nH_{2n-2} (d) C_nH_{3n-2}

29. The correct structure for hexene is:

(a) C—C—C—C—C—C
(b) C—C—C—C—C=C
(c) C—C—C—C—C=C—C
(d) C—C—C—C=C

30. Which of the following is the organic compound?

(a) CO (b) CO_2 (c) Na_2CO_3 (d) C_2H_6

31. The correct structure for propene is:

(a) C—C—C
(b) C—C=C
(c) C—C=C—C
(d) C—C—C=C

32. Which of the following is not considered an organic compound?

(a) CH_3Cl (b) CH_3OH (c) Na_2CO_3 (d) C_2H_6

33. The bond angle for the carbon-hydrogen bonds in a molecule of methane (CH_4) is approximately:
(a) 90° (b) 45° (c) 109° (d) 180°

34. The correct name for C—C—C—C—C—C=C is:
(a) hexene (b) hexyne
(c) heptyne (d) heptene

35. The smallest cyclic alkane would have the formula:
(a) C_3H_6 (b) C_2H_6 (c) C_3H_8 (d) C_4H_8

36. The correct name for C—C—C—C≡C is:
(a) hexene (b) hexyne
(c) pentyne (d) pentane

37. A cyclic alkane that has five carbon atoms would also contain:
(a) 12 hydrogen atoms (b) 10 hydrogen atoms
(c) 8 hydrogen atoms (d) 6 hydrogen atoms

CHAPTER 19 Quiz Answers

1. Graphite and diamond

2. C—C—C—C—C   C—C—C(—C)—C   C—C(—C)(—C)—C

3. (a) C_nH_{2n+2} (b) C_nH_{2n} (c) C_nH_{2n-2}

4. (a) 2,3-dimethylhexane (b) 3-hexene
(c) 3-ethyl-1-hexene (d) propylcyclopentane

5. (a) C—C—C—C—C(—C)—C—C

(b)

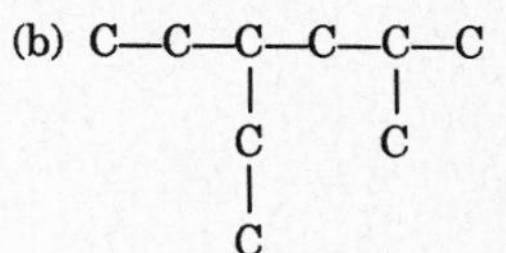

(c)

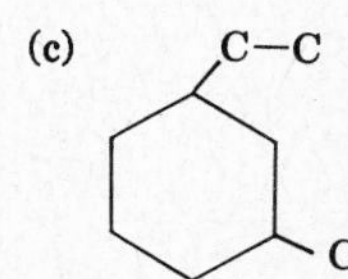

(d) C—C(—C)—C=C

6. False

7. True


```
                                          C
                                          |
8.  C—C—C—C—Cl     C—C—C—C            C—C—C
                         |                |
                         Cl               Cl
```

9. (1) b, (2) a, (3) c

```
10. CH3—CH=CH—CH2—CH—C≡CH
                  |
                  CH2
                  |
                  CH3
```

```
11. (a) C—C—C                (b) C—C—C—C—C—C
          |                              | |
          C                              C C
                                         |
                                         C
```

12. Ethene, propene, butene, pentene, hexene

13. (a) 2-pentene (b) 2,5-dimethyl-3-heptene

14. (a) 4,5,5-trimethyl-2-hexyne (b) 2,6-dimethyl-3-octyne

15. (a) Cyclohexane (b) Methylcyclopentane
    (c) Cyclohexene (d) 1,3-dimethylcyclohexane

```
                    C                    C
                    |                    |
16. (a) C—C—C—C—C—C          (b) C—C—C—C—C
                  | |                    |
                  C C                    C
                                         |
                                         C
```

```
                                         C
                                         |
17. (a) C—C—C=C—C            (b) C—C—C—C—C=C
              |                          |
              C                          C
              |                          |
              C                          C
```

```
                    C
                    |
18. (a) C—C—C—C—C—C—C—C≡C        (b) C—C—C—C≡C
                    |   |                  |
                    C   C                  C
                                           |
                                           C
```


19. (a)

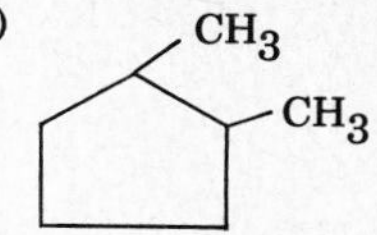

(b) 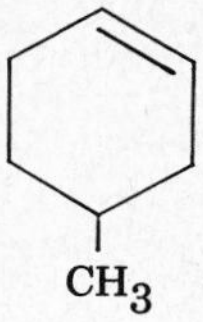 CH_3

20. (a)

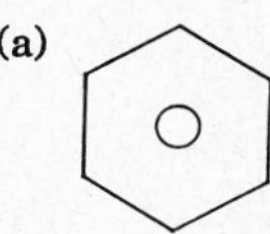

(b) CH_3

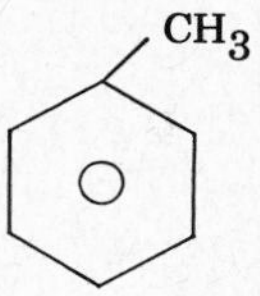

21. d	22. d	23. c	24. d	25. b
26. c	27. b	28. a	29. b	30. d
31. b	32. c	33. c	34. d	35. a
36. c	37. b			

CHAPTER 20 Quiz

1. Determine the class of each of the following compounds:

   (a) $CH_3CH_2—O—CH_2CH_2CH_3$

   (b) $CH_3CH_2CH_2—OH$

   (c) $CH_3CH_2—NH_2$

   (d) $CH_3CH_2—C(=O)—CH_3$

   (e) $CH_3CH_2—C(=O)—OH$

   (f) $CH_3CH_2—C(=O)—O—CH_3$

2. Name the following compounds, using either their common or their IUPAC names:

   (a) $CH_3CH_2CH_2—OH$

   (b) $CH_3—C(=O)—CH_3$

   (c) $CH_3CH_2—C(=O)—H$

   (d) $CH_3CH_2—C(=O)—O—CH_3$

3. Write the formula for each of the following compounds:

   (a) Butanoic acid
   (b) Diethyl ether
   (c) 2-aminopropane
   (d) Cyclohexanol

4. Complete the following reaction:

   $CH_3CH_2—C(=O)—OH + HO—CH_3 \rightarrow$

5. For each of the following alcohols, state which will dissolve in water and which will not.

   (a) CH_3CH_2OH

   (b) $CH_3—CH_2—CH_2—CH_2—CH_2—CH_2—CH_2—OH$

6. Match each word on the left with its description on the right.

(a) Alcohol	(1) R—C(=O)—R′
(b) Aldehyde	(2) R—OH
(c) Ketone	(3) R—O—R′
(d) Ether	(4) R—C(=O)—H

7. Name each of the following alcohols:

   (a) $CH_3CH_2CH_2CH_2—OH$ (b) CH_3CH_2OH

8. Determine whether each of the following alcohols is polar or nonpolar.

   (a) $CH_3CH_2CH_2CH_2CH_2CH_2CH_2—OH$

   (b) $CH_3CH_2—OH$

9. Write the structure of each of the following alcohols.

   (a) 4-heptanol (b) 2-hexanol

10. Name each of the following ethers:

    (a) $CH_3CH_2CH_2—O—CH_2CH_2CH_3$

    (b) $CH_3CH_2—O—CH_2CH_2CH_2CH_3$

11. Write the structure of each of the following ethers:

    (a) Dibutyl ether (b) Methyl propyl ether

12. Name each of the following aldehydes:

    (a) $CH_3—CH_2—CH_2—CH_2—C(=O)—H$

    (b) $CH_3—CH_2—CH_2—CH_2—CH_2—CH_2—C(=O)—H$

13. Name each of the following ketones:

    (a) 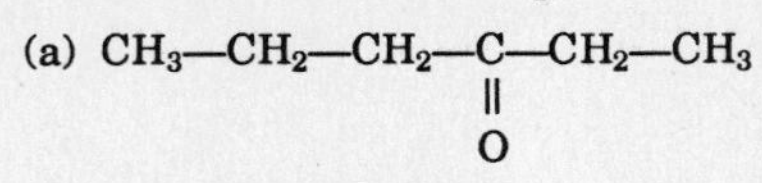

    (b) 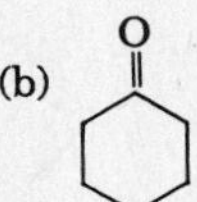

14. Write the structure of each of the following ketones:

    (a) Dimethylketone (b) 3-octanone

15. Name each of the following carboxylic acids:

(a) $CH_3—CH_2—CH_2—CH_2—CH_2—\underset{\underset{O}{\|}}{C}—OH$

(b) $H—\underset{\underset{O}{\|}}{C}—OH$

16. Write the structure of each of the following carboxylic acids:

(a) Ethanoic acid

(b) Propionic acid

17. Write the product of the following esterification reaction:

$CH_3CH_2CH_2CH_2—\underset{\underset{O}{\|}}{C}—OH + HO—CH_3 \xrightarrow{H^{1+}} ?$

18. Name each of the following esters:

(a) $CH_3CH_2—\underset{\underset{O}{\|}}{C}—OCH_2CH_3$

(b) $CH_3CH_2CH_2—\underset{\underset{O}{\|}}{C}—OCH_3$

19. Name each of the following amines:

(a) $CH_3—\underset{\underset{CH_3}{|}}{N}—CH_3$

(b) $CH_3CH_2CH_2—NH_2$

20. Write the structure of each of the following amines:

(a) Ethylpropyl amine

(b) 3-aminohexane

21. The general formula for an ester is:

(a) $R—\underset{\underset{O}{\|}}{C}—R'$

(b) $R—\underset{\underset{O}{\|}}{C}—O—R'$

(c) $R—O—R'$

(d) $R—\underset{\underset{O}{\|}}{C}—O—H$

22. The general formula for an ether is:

(a) $R—\underset{\underset{O}{\|}}{C}—O—R'$

(b) $R—O—R'$

(c) $R—\underset{\underset{O}{\|}}{C}—R'$

(d) $R—OH$

23. The general formula for a carboxylic acid is:

(a) $R—\underset{\underset{O}{\|}}{C}—O—H$

(b) $R—O—R'$

(c) $R—\underset{\underset{O}{\|}}{C}—R$

(d) $R—OH$

24. The general formula for a ketone is:

(a) R—C(=O)—R′

(b) 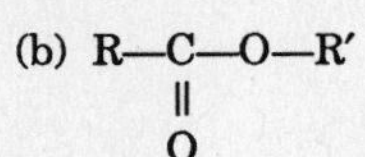

(c) R—O—R′

(d) 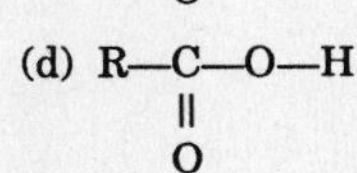

25. The general formula for an alcohol is:

(a) R—OH

(b) 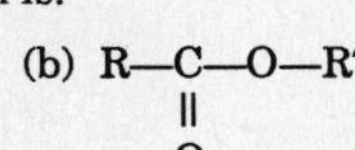

(c) R—O—R′

(d) 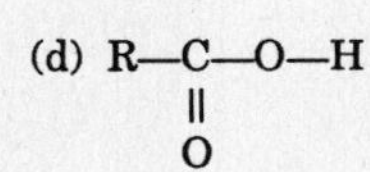

26. The correct structure for ethanol is:

(a) CH_3—O—CH_3

(b) CH_3—CH_2—OH

(c) CH_3—C(=O)—CH_3

(d) CH_3—C(=O)—H

27. The correct structure for butanal is:

(a) CH_3CH_2—O—CH_2CH_3

(b) $CH_3CH_2CH_2CH_2OH$

(c) $CH_3CH_2CH_2C$(=O)—OH

(d) $CH_3CH_2CH_2C$(=O)—H

28. The correct structure for 2-pentanone is:

(a) CH_3CH_2—O—$CH_2CH_2CH_3$

(b) $CH_3CH_2CH_2CH_2CH_2OH$

(c) $CH_3CH_2CH_2C$(=O)CH_3

(d) $CH_3CH_2CH_2CH_2C$(=O)—H

29. The correct structure for n-propanol is:

(a) $CH_3CH_2CH_2$—OH

(b) $CH_3CH_2CH_2$—O—$CH_2CH_2CH_3$

(c) CH_3CH_2CH
  ‖
  O

(d) CH_3CH_2C—OH
  ‖
  O

30. RNH_2 is the general formula for a(n):
(a) alcohol (b) acid (c) ester (d) amine

CHAPTER 20 Quiz Answers

1. (a) Ether (b) Alcohol (c) Amine (d) Ketone
   (e) Acid (f) Ester

2. (a) 1-propanol or n-propyl alcohol
   (b) Acetone or dimethyl ketone or propanone
   (c) Propanal
   (d) Methylpropanoate

3. (a) $CH_2CH_2CH_2$—C—OH
  ‖
  O

   (b) CH_3CH_2—O—CH_2CH_3

   (c) CH_3CHCH_3
  |
  NH_2

   (d) 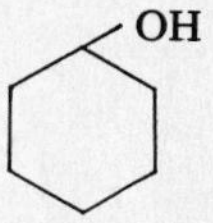 OH (on cyclohexane ring)

4. CH_3CH_2—C—O—CH_3 + H_2O (the products)
  ‖
  O

5. (a) Soluble in water (b) Insoluble in water

6. (1) c (2) a (3) d (4) b

7. (a) 1-butanol (b) Ethanol

8. (a) Nonpolar (b) Polar

9. (a) $CH_3CH_2CH_2$—CH—CH_2CH_2—CH_3
  |
  OH

   (b) CH_3—CH—$CH_2CH_2CH_2CH_3$
  |
  OH

10. (a) Dipropyl ether (b) Butyl ethyl ether

11. (a) $CH_3CH_2CH_2CH_2{-}O{-}CH_2CH_2CH_2CH_3$
    (b) $CH_3{-}O{-}CH_2CH_2CH_3$

12. (a) Pentanal (b) Heptanal

13. (a) 3-hexanone (b) Cyclohexanone

14. (a) $CH_3{-}\underset{\underset{\displaystyle O}{\|}}{C}{-}CH_3$

    (b) $CH_3CH_2{-}\underset{\underset{\displaystyle O}{\|}}{C}{-}CH_2CH_2CH_2CH_2CH_3$

15. (a) Hexanoic acid (b) Methanoic acid

16. (a) $CH_3\underset{\underset{\displaystyle O}{\|}}{C}{-}OH$ (b) $CH_3CH_2\underset{\underset{\displaystyle O}{\|}}{C}{-}OH$

17. $CH_3CH_2CH_2CH_2{-}\underset{\underset{\displaystyle O}{\|}}{C}{-}O{-}CH_3$

18. (a) Ethylpropanoate (b) Methylbutanoate

19. (a) Trimethyl amine (b) Propyl amine

20. (a) $CH_3CH_2{-}\underset{\underset{\displaystyle H}{|}}{N}{-}CH_2CH_2CH_3$

    (b) $CH_3CH_2{-}\underset{\underset{\displaystyle NH_2}{|}}{CH}{-}CH_2CH_2CH_3$

21. b 22. b 23. a 24. a 25. a

26. b 27. d 28. c 29. a 30. d

PART III

FINAL EXAMINATIONS WITH ANSWERS

PART III

FINAL EXAMINATION I

CHAPTER 1 Questions

1. The modern age of chemistry dawned with the publication of Boyle's book,

   (a) The End of Alchemy
   (b) Modern Chemistry
   (c) The Sceptical Chymist
   (d) Basic Concepts of Chemistry

2. Which of the following is not one of the four basic elements that Empedocles thought made up the world?

   (a) earth (b) oxygen (c) fire (d) air

CHAPTER 2 Questions

3. The formula for density is:

   (a) $D = \frac{m}{V}$ (b) $D = m \times V$ (c) $D = \frac{V}{m}$ (d) $D = \frac{m}{V^2}$

4. If 1.00 meter equals 3.28 feet, then 5.50 meters equals:

   (a) 0.055 feet (b) 0.60 feet
   (c) 18.0 feet (d) 1.66 feet

5. A mass of 0.430 g is equivalent to:

   (a) 43.0 mg (b) 430 kg (c) 4.3 cg (d) 430 mg

6. A temperature of 122°F is equivalent to:

   (a) 50.0°C (b) 67.7°C (c) $1\overline{00}$°C (d) 188°C

7. A mass of a cube that measures 4.00 cm on a side and has a density of 1.20 g/cm^3 is:

   (a) 53.3 g (b) 16 g (c) 4.96 g (d) 76.8 g

CHAPTER 3 Questions

8. Which of the following is not an example of a chemical change?

(a) burning paper (b) cooking an egg
(c) breaking glass (d) toasting bread

9. Which of the following is an element?

(a) potassium (b) carbon dioxide
(c) water (d) salt

10. The symbol Hf stands for:

(a) the compound hydrogen fluoride
(b) the element hafnium
(c) a mixture of hydrogen and fluorine
(d) a solution of hydrogen ion and fluoride ion

11. The fact that water is always H_2O is a statement of the:

(a) Law of Conservation of Mass
(b) Ideal Gas Law
(c) Henry's Law
(d) Law of Definite Proportions

12. The smallest particle of matter that can enter into chemical combination is a(n):

(a) atom (b) molecule
(c) mixture (d) solution

CHAPTER 4 Questions

13. The scientist responsible for the "energy-level atom" is

(a) Dalton (b) Thomson
(c) Rutherford (d) Bohr

14. Isotopes of the same element have:

(a) the same number of protons
(b) the same number of neutrons
(c) different number of protons
(d) the same atomic mass

15. The isotope $^{238}_{92}U$ has:

(a) 92p, 92e, 146n (b) 238p, 238e, 92n
(c) 92p, 238e, 238n (d) 92p, 92e, 238n

CHAPTER 5 Questions

16. The maximum number of electrons permitted in the M energy level is:

(a) 2 (b) 8 (c) 18 (d) 50

17. What is the number of electrons in each of the energy levels of a sulfur atom?

(a) K has 2, L has 14.
(b) K has 2, L has 8, M has 6.
(c) K has 2, L has 8, M has 8, N has 14.
(d) K has 2, L has 18, M has 2.

CHAPTER 6 Questions

18. The electron configuration of $_{15}P$ is:

(a) $1s^22s^22p^63s^23p^3$ (b) $1s^22s^22p^{10}3p^1$
(c) $1s^22s^22p^83p^4$ (d) $1s^22s^22p^63s^23p^63d^{10}4s^24p^1$

19. Of the four elements listed below, the element with the largest atomic radius is:

(a) K (b) Li (c) Rb (d) Na

20. The least reactive elements are found in Group

(a) IA (b) VIIIA (c) IIA (d) IVA

CHAPTER 7 Questions

21. The correct bonding for CO_2 is:

(a) O=C=O (b) O—C—O
(c) O=C—O (d) C—O—O

22. Which compound has a nonpolar covalent bond?

(a) H_2O (b) NaF (c) H_2S (d) Cl_2

CHAPTER 8 Questions

23. The correct formula for iron(III) oxide is:

(a) FeO (b) Fe_2O_3 (c) Fe_3O_2 (d) Fe_2O

24. The correct formula for ammonium sulfate is:

(a) NOS (b) $(NH_4)_2S$ (c) $(NH_4)_2SO_4$ (d) $NH_4(SO_4)_2$

25. The correct name for $Cu_3(PO_4)_2$ is:

(a) copper(I) phosphide
(b) copper(II) phosphate
(c) copper(III) phosphate
(d) copper phosphate

CHAPTER 9 Questions

26. A sample of SO_2 that weighs 3.2 g contains:

(a) 1.0 mole of SO_2 molecules
(b) 0.50 mole of SO_2 molecules
(c) 20 moles of SO_2 molecules
(d) 0.05 mole of SO_2 molecules

27. The empirical formula of a compound that has 40.0 percent C, 6.60 percent H, and 53.4 percent O is:

(a) C_2HO
(b) CH_2O
(c) CHO_2
(d) C_2HO_2

28. A substance has a molecular weight of 56 and an empirical formula of CH_2. What is its molecular formula?

(a) CH_2
(b) C_4H_8
(c) C_2H_4
(d) C_3H_6

CHAPTER 10 Questions

29. In the equation $N_2 + H_2 \rightarrow NH_3$ (unbalanced), the sum of the coefficients is:

(a) 3
(b) 4
(c) 5
(d) 6

(Hint: After balancing the equation, don't forget to include coefficients that are the number 1.)

30. The product of the reaction $H_2 + Cl_2$ is:

(a) H_2Cl_2
(b) HCl
(c) HCl_2
(d) H_2Cl

31. In the reaction $Zn + 2HCl \rightarrow ZnCl_2 + H_2$, the substance that is oxidized is:

(a) Zn
(b) H
(c) Cl
(d) Nothing is oxidized.

CHAPTER 11 Questions

32. How many grams of CO_2 can be produced from the burning of 22 g of propane (C_3H_8)?

$C_3H_8 + 5O_2 \rightarrow 3CO_2 + 4H_2O$

(a) 33 g
(b) 99 g
(c) 66 g
(d) 132 g

33. How many grams of water can be produced from 64 g of oxygen and $1\overline{0}$g of hydrogen?

    (a) 74 g (b) 72 g (c) 90 g (d) 36 g

CHAPTER 12 Questions

34. The ΔH_R for the combustion of 1.00 mole of methane (CH_4) is:

    (a) –230.7 kcal (b) –68.3 kcal
    (c) –17.9 kcal (d) –212.8 kcal

    ΔH_f for $CH_4(g) = -17.9$ kcal, ΔH_f for $CO_2(g) = -94.1$ kcal,

    ΔH_f for $H_2O(\ell) = -68.3$ kcal

35. How does the sum of the heat contents of the products compare with those of the reactants in an exothermic reaction?

    (a) The products have greater heat content than the reactants.
    (b) The products have lower heat content than the reactants.
    (c) The products have the same heat content as the reactants.
    (d) Heat content can vary from reaction to reaction.

CHAPTER 13 Questions

36. A pressure of 1520 torr is equivalent to:

    (a) 1.00 atm (b) 2.00 atm (c) 0.500 atm (d) 14.7 lb/in^2

37. A gas has a volume of 50.0 mL at a pressure of 1.00 atm. What will the volume of the gas at the same temperature be if the pressure is increased to 2.00 atm?

    (a) 25.0 mL (b) 50.0 mL (c) $1\overline{00}$ mL (d) 10.0 mL

38. When the Kelvin temperature of a gas is doubled at constant pressure, its volume

    (a) is reduced by one-half (b) is reduced by one-fourth
    (c) is doubled (d) is tripled

39. A gas has a mass of 1.00 g at a pressure of 1.50 atm and a temperature of $3\overline{00}$°K. Its volume is 82.0 mL. What is its molecular weight?

    (a) 50.0 g/mole (b) $1\overline{00}$ g/mole
    (c) $2\overline{00}$ g/mole (d) $4\overline{00}$ g/mole

CHAPTER 14 Questions

40. The boiling point of a substance ____________ with increasing atmospheric pressure. (Fill in the blank with one of the choices that follow.)

(a) decreases
(b) increases
(c) stays the same
(d) Insufficient information to answer this question

41. A certain compound has a high melting point, does not conduct electricity, and is brittle. This type of solid is:

(a) ionic
(b) molecular
(c) atomic
(d) None of these answers

42. A substance that has definite volume but indefinite shape is:

(a) a solid
(b) a liquid
(c) a gas
(d) None of these answers

CHAPTER 15 Questions

43. A solution prepared by dissolving 20.0 g of salt in 60.0 g of water has a percentage-by-weight concentration of

(a) 33.3 percent
(b) 20.0 percent
(c) 25.0 percent
(d) 50.0 percent

44. The molarity of a solution prepared by dissolving 72 g of HCl in sufficient water to make 0.50 liter of solution is:

(a) 1.0M (b) 2.0M (c) 3.0M (d) 3.9M

45. The normality of a 0.5M $Ca(OH)_2$ solution is:

(a) 0.25N (b) 1N (c) 2N (d) 4N

CHAPTER 16 Questions

46. The formula for phosphoric acid is:

(a) H_3P (b) H_3PO_4 (c) HPO_4 (d) H_2PO_3

47. The pH of a 0.001M NaOH solution is:

(a) 3 (b) 1 (c) 13 (d) 11

48. The NH_4^{1+} ion is a Brønsted-Lowry:

(a) acid (b) base (c) salt (d) oxide

CHAPTER 17 Questions

49. The K_{eq} expression for the reaction $2A(g) + 3B(g) = 5C(g) + 4D(g)$ is:

(a) $K_{eq} = \frac{[A]^2[B]^3}{[C]^5[D]^4}$ (b) $K_{eq} = \frac{[2A][3B]}{[5C][4D]}$

(c) $K_{eq} = \frac{[C]^5[D]^4}{[A]^2[B]^3}$ (d) $K_{eq} = \frac{[5A][4D]}{[2A][3B]}$

50. Which of the following statements is true?

(a) In a reaction at equilibrium, the concentrations of the reactants and products are the same.
(b) A catalyst increases reaction rates by changing reaction mechanisms.
(c) The amount of product obtained at equilibrium is proportional to how fast equilibrium is obtained.
(d) A catalyst has no effect on the activation energy of a reaction.

51. For the reaction $2NO(g) + O_2(g) = 2NO_2(g)$ at a given temperature, the equilibrium concentrations are as follows:

$NO(g)$ = 0.600 mole/liter, $O_2(g)$ = 0.800 mole/liter, and $NO_2(g)$ = 4.40 moles/liter.

The K_{eq} for this reaction is

(a) 67.2 (b) 1.5×10^{-2} (c) 9.17 (d) 5.1×10^{-1}

52. The equilibrium concentration of OH^{-1} in a 0.100M ammonium hydroxide solution is:

(a) 1.8×10^{-5} (b) 1.8×10^{-6}

(c) 4.24×10^{-3} (d) 1.34×10^{-3}

(Note: K_b for ammonium hydroxide is 1.80×10^{-5}.)

53. If the concentrations of Ba^{+2} and SO_4^{-2} ions are each 3.9×10^{-5} mole/liter at equilibrium at 25°C, the K_{sp} value for $BaSO_4$ at this temperature is:

(a) 3.9×10^{-5} (b) 1.5×10^{-9}

(c) 6.7×10^{8} (d) 3.0×10^{-9}

CHAPTER 18 Questions

54. A substance has a half-life of one day. You have $5\overline{00}$ g of this substance today. How many grams will you have at the end of four days?

(a) 250 g (b) 125 g (c) 62.5 g (d) 50.0 g

55. An air-pollution control device used to remove water-soluble gases is known as:

(a) a scrubber (b) an electrostatic precipitator
(c) an ion-exchange column (d) an air purifier

56. A device that produces electrical energy from chemical energy is known as:

(a) a voltaic cell (b) an electrolytic cell
(c) a heat engine (d) None of these things

CHAPTER 19 Questions

57. In organic compounds, carbon usually forms:

(a) one bond (b) two bonds
(c) three bonds (d) four bonds

58. Which of the following hydrocarbons is not a structural isomer of the others?

```
(a) C—C—C—C—C            (b) C—C—C—C—C—C
       |
       C

(c)    C                 (d)    C
       |                        |
    C—C—C—C                  C—C—C
       |                        |
       C                        C
```

59. The name of the hydrocarbon C—C—C=C is:

```
C—C—C=C
  |
  C
```

(a) methylbutane (b) 2-methylbutene
(c) 3-methyl-1-butene (d) 2-methyl-1-butene

CHAPTER 20 Questions

60. The general formula of an ester is:

```
(a) R—C—R'               (b) R—C—O—R'
      ||                       ||
      O                        O

(c) R—O—R                (d) R—C—O—H
                               |
                               O
```

61. The name of the compound CH_3CH_2—C—H is:

```
CH3CH2—C—H
       ||
       O
```

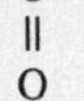

(a) propanal (b) propanol
(c) propanoic acid (d) ethyl methanoate

62. The structural formula of 3-heptanone is:

(a) $CH_3CH_2CH_2CH_2CH_2CH_2{-}\underset{\underset{O}{\|}}{C}{-}H$

(b) $CH_3CH_2CH_2{-}O{-}CH_2CH_2CH_2CH_3$

(c) $CH_3CH_2{-}C{-}O{-}CH_2CH_2CH_3$

(d) $CH_3CH_2{-}\underset{\underset{O}{\|}}{C}{-}CH_2CH_2CH_2CH_3$

FINAL EXAMINATION I: Answer Key

1. c	2. b	3. a	4. c	5. d
6. a	7. d	8. c	9. a	10. b
11. d	12. a	13. d	14. a	15. a
16. c	17. b	18. a	19. c	20. b
21. a	22. d	23. b	24. c	25. b
26. d	27. b	28. b	29. d	30. b
31. a	32. c	33. b	34. d	35. b
36. b	37. a	38. c	39. c	40. b
41. a	42. b	43. c	44. d	45. b
46. b	47. d	48. a	49. c	50. b
51. a	52. d	53. b	54. c	55. a
56. a	57. d	58. d	59. c	60. b
61. a	62. d			

FINAL EXAMINATION II

CHAPTER 1 Questions

1. Which of the following is one of the four basic elements that Empedocles thought made up the world?

   (a) gold (b) silver (c) earth (d) mercury

2. The major aim of the alchemists was:

   (a) to make embalming potions
   (b) to turn lead into gold
   (c) to make organic acids
   (d) to make plastics

CHAPTER 2 Questions

3. Which of the following is not a unit of density?

   (a) g/cm^3 (b) g/liter
   (c) pounds/gallon (d) pounds/square inch

4. If 1.00 inch equals 2.54 cm, then 12.0 inches equals:

   (a) 30.5 cm (b) 4.72 cm (c) 0.212 cm (d) 24.0 cm

5. A mass of 2.25 kg is equivalent to:

   (a) 22.5 g (b) 2250 mg (c) 2250 g (d) 2250 dg

6. A temperature of 20.0°C is equivalent to:

   (a) 68.0°F (b) 20.0°F (c) 6.67°F (d) 100°F

CHAPTER 3 Questions

7. Which of the following is not a compound?

   (a) glucose (b) calcium
   (c) water (d) sodium chloride

8. What is the total number of atoms in one molecule of $C_{13}H_{14}N_4O_3S$?
   (a) 1 (b) 5 (c) 35 (d) 13

9. In the periodic table, if oxygen were assigned a mass of 1, what would be the atomic mass of bromine?
   (a) 2 (b) 3 (c) 4 (d) 5

CHAPTER 4 Questions

10. An isotope that has 35p, 35e, and 46n is:
    (a) $^{81}_{46}Pd$ (b) $^{81}_{35}Br$ (c) $^{46}_{35}Br$ (d) $^{35}_{46}Pb$

11. The scientist responsible for the plum-pudding model of the atom was:
    (a) Dalton (b) Rutherford
    (c) Thomson (d) Bohr

12. Isotopes of the same element always have:
    (a) the same number of neutrons
    (b) the same atomic weight
    (c) different numbers of protons
    (d) the same atomic number

CHAPTER 5 Questions

13. The maximum number of electrons permitted in the L energy level is:
    (a) 2 (b) 8 (c) 18 (d) 32

14. The group number of an element that has two electrons in the K level and three electrons in the L level is:
    (a) IA (b) IIA (c) IIIA (d) IVA

CHAPTER 6 Questions

15. The maximum number of electrons allowed in the d sublevel is:
    (a) 2 (b) 6 (c) 10 (d) 8

16. Of the four elements listed below, the element with the smallest atomic radius is:
    (a) K (b) Li (c) Rb (d) Na

17. Of the four elements listed below, the element with the lowest ionization potential is:

(a) Mg (b) Ca (c) Sr (d) Ba

CHAPTERS 7 and 8 Questions

18. Which of the following compounds is the least polar?

(a) HF (b) HCl (c) HBr (d) HI

19. The formula of copper(I) sulfate is:

(a) $CuSO_4$ (b) $Cu(SO_4)_2$ (c) Cu_2SO_4 (d) $CuSO_3$

20. The name of the compound $Fe(C_2H_3O_2)_3$ is:

(a) iron(I) acetate (b) iron(III) acetate
(c) iron(III) carbonate (d) iron(I) carbonate

21. Which of the following molecules is polar?

(a)
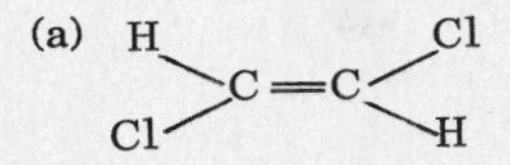

(b)
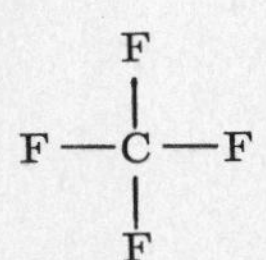

(c) N≡N

(d)
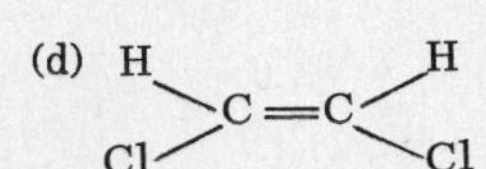

22. Which of the following bonds is nonpolar?

(a) H_2S (b) HCl (c) Br_2 (d) OF_2

CHAPTER 9 Questions

23. How many moles of molecules are there in 198 g of CO_2?

(a) 2.00 moles (b) 2.50 moles
(c) 4.00 moles (d) 4.50 moles

24. The empirical formula of a compound is CH, and its molecular mass is 78.0. What is its molecular formula?

(a) CH_4 (b) C_3H_3 (c) C_6H_6 (d) C_6H_{12}

CHAPTER 10 Questions

25. Balance the following equation:

    _____$KClO_3 \rightarrow$ _____$KCl +$ _____O_2

    The sum of the coefficients is:

    (a) 4 (b) 5 (c) 6 (d) 7

26. Identify the substance being oxidized in the following equation:

    $2Al + Fe_2O_3 \rightarrow Al_2O_3 + 2Fe$

    (a) aluminum (b) iron(III) ion
    (c) oxide ion (d) iron metal

27. For the reaction in Question 26, the substance being reduced is:

    (a) aluminum (b) iron(III) ion
    (c) oxide ion (d) iron metal

28. The reaction in Question 26 can also be classified as:

    (a) a combination reaction
    (b) a decomposition reaction
    (c) a single-replacement reaction
    (d) a double-replacement reaction

29. The product of the reaction $Hg + O_2 \rightarrow$ is:

    (a) Hg_2O (b) HgO (c) HgO_3 (d) HgO_2

CHAPTER 11 Questions

30. How many grams of H_2O can be produced from 128 g of O_2 and 12.0 g of H_2?

    $2H_2 + O_2 \rightarrow 2H_2O$

    (a) 108 g (b) 64.0 g (c) 36.0 g (d) 72.0 g

31. According to the equation $N_2 + 3H_2 \rightarrow 2NH_3$, how many moles of ammonia can be produced from 8.0 moles of hydrogen?

    (a) 7.5 moles (b) 12 moles
    (c) 6 moles (d) 5.3 moles

CHAPTER 12 Questions

32. The ΔH_f for CH_4 is –17.9 kcal/mole. How much heat is produced from the formation of 64.0 g of CH_4?

    (a) 17.9 kcal (b) 35.8 kcal (c) 53.7 kcal (d) 71.6 kcal

33. How does the sum of the heat contents of the products compare with that of the reactants in an endothermic reaction?
    (a) The products have greater heat content than the reactants.
    (b) The products have less heat content than the reactants.
    (c) The products have the same heat content as the reactants.
    (d) Heat content can vary from reaction to reaction.

CHAPTER 13 Questions

34. When the temperature of a gas is increased at constant pressure, its volume:
    (a) increases
    (b) decreases
    (c) stays the same
    (d) Insufficient information to answer this question

35. A gas has a volume of 6.0 liters at a pressure of 380 torr. The pressure is increased to 760 torr at the same temperature. What is its new volume?
    (a) 6.0 liters (b) 3.0 liters (c) 12 liters (d) 9.0 liters

36. When the pressure of a gas is increased at constant temperature, its volume:
    (a) increases
    (b) decreases
    (c) stays the same
    (d) Insufficient information to answer this question

37. What is the volume of 0.500 mole of O_2 gas at STP?
    (a) 5.60 liters (b) 22.4 liters
    (c) 11.2 liters (d) 44.8 liters

CHAPTER 14 Questions

38. Which of the following is not a true statement about solids?
    (a) The melting point of a solid is not affected by atmospheric pressure.
    (b) Solids have definite shape.
    (c) Solids have definite volume.
    (d) Solids are easily compressible.

39. A substance that has indefinite volume and indefinite shape is:
    (a) a solid (b) a liquid (c) a gas (d) All of these answers

40. The boiling point of a substance ______________ with decreasing atmospheric pressure. (Fill in the blank with one of the following choices.)

(a) increases
(b) decreases
(c) stays the same
(d) Insufficient information to answer this question

CHAPTER 15 Questions

41. What is the molarity of a solution prepared by dissolving 18.0 g of glucose in enough water to make $2\bar{0}\bar{0}$ mL of solution? The molecular mass of glucose is $18\bar{0}$ g/mole.

(a) 0.0900M (b) 2.00M (c) 0.500M (d) 1.00M

42. What is the percentage of mass volume of a solution prepared by dissolving 25 g of sodium chloride in enough water to make 125 mL of solution?

(a) $2\bar{0}$ percent (b) $5\bar{0}$ percent
(c) 5.0 percent (d) 25 percent

43. The normality of a 0.50M H_3PO_4 solution is:

(a) 0.50N (b) 1.0N (c) 2.0N (d) 1.5N

CHAPTER 16 Questions

44. The formula for nitric acid is:

(a) H_3PO_4 (b) HNO_3 (c) HON (d) H_2CO_3

45. The pH of a substance is 4. Its hydrogen-ion concentration is:

(a) 0.01M (b) 0.1M (c) 0.0001M (d) 0.001M

46. The pH of a substance is 8. Its pOH is:
(a) 8 (b) 2 (c) 10 (d) 6

47. In the reaction $NH_3 + H_2O \rightarrow NH_4^{1+} + OH^{1-}$, the NH_3 is a Brønsted-Lowry:

(a) acid (b) base (c) salt (d) oxide

CHAPTER 17 Questions

48. The K_{eq} expression for the reaction $H_2(g) + I_2(g) = 2HI(g)$ is

(a) $K_{eq} = \frac{[H_2][I_2]}{[HI]^2}$

(b) $K_{eq} = \frac{[H_2]^2[I_2]^2}{[HI]}$

(c) $K_{eq} = \frac{[HI]^2}{[H_2][I_2]}$

(d) $K_{eq} = [H_2][I_2]$

49. If the $NH_3(g)$ concentration is increased by adding $NH_3(g)$ to the reaction

$N_2(g) + 3H_2(g) = 2NH_3(g)$

which is at equilibrium, the equilibrium will shift to the:

(a) left
(b) right
(c) no shift
(d) Insufficient information to answer this question

50. If the equilibrium concentrations are $HC_7H_5O_2$ = 0.0975 mole/liter, H^{1+} = 0.0025 mole/liter, and $C_7H_5O_2^{1-}$ = 0.0025 mole/liter, the K_a for benzoic acid at 25°C is __________.

(a) 1.56×10^4

(b) 6.4×10^{-5}

(c) 6.2×10^{-6}

(d) 2.56×10^{-2}

51. The solubility of CuS at 25°C in grams per liter is __________, given that the K_{sp} is 9.0×10^{-45}.

(a) 9.0×10^{-45} g/liter

(b) 9.5×10^{-23} g/liter

(c) 3×10^{23} g/liter

(d) 9.1×10^{-21} g/liter

52. For the equilibrium

$H_2O + CO_2 = H_2CO_3 = H^{1+} + HCO_3^{1-}$

removal of H^{1+} ion from the system shifts the equilibrium to the:

(a) left
(b) right
(c) no change
(d) Insufficient information to answer this question

CHAPTER 18 Questions

53. A beta particle can be represented as:

(a) ${}^{0}_{-1}e$ (b) ${}^{4}_{2}He$ (c) ${}^{0}_{1}n$ (d) ${}^{1}_{1}p$

54. The isotope I-131 has a half-life of eight days. You have 40.0 g of I-131 today. How many grams will you have left in 24 days?

(a) 5.00 g (b) 10.0 g (c) 15.0 g (d) 20.0 g

55. An air-pollution-control device used to remove small particles of dust is known as:

(a) a scrubber
(b) an electrostatic precipitator
(c) an ion-exchange column
(d) an air purifier

CHAPTER 19 Questions

56. In a carbon-carbon triple bond, the two tetrahedral carbon atoms share a common:

(a) point
(b) edge
(c) side
(d) None of these answers

57. Which of the following compounds is not a structural isomer of the others?

(a) C—C—C—C—C

(b)
```
      C
      |
   C—C—C—C
      |
      C
```

(c)
```
      C
      |
   C—C—C
      |
      C
```

(d)
```
      C
      |
   C—C—C—C—C
      |
      C
```

58. The name of the compound C—C=C—C—C is:
```
C—C=C—C—C
        |
        C
```

(a) 2-methyl-3-pentene
(b) 4-methyl-2-pentene
(c) 2-methyl-2-pentene
(d) 4-methyl-4-pentene

CHAPTER 20 Questions

59. The general formula for an ether is:

(a)
```
R—C—O—R'
  ||
  O
```

(b) R—O—R'

(c)
```
R—C—R'
  ||
  O
```

(d) R—OH

60. The name of the compound $CH_3—CH_2—CH_2—C—OH$ is:
```
CH3—CH2—CH2—C—OH
            ||
            O
```

(a) butanol
(b) butanal
(c) butanoic acid
(d) butanone

61. The structure of 2-methylpentanal is:

(a) $CH_3\text{—}CH_2\text{—}CH_2\text{—}CH(CH_3)\text{—}C(=O)\text{—}H$

(b) $CH_3\text{—}CH(CH_3)\text{—}CH_2\text{—}CH_2\text{—}C(=O)\text{—}H$

(c) $CH_3\text{—}CH_2\text{—}CH_2\text{—}CH_2\text{—}C(=O)\text{—}O\text{—}CH_3$

(d) $CH_3\text{—}CH_2\text{—}C(=O)\text{—}CH(CH_3)\text{—}CH_3$

FINAL EXAMINATION II: Answer Key

1. c	2. b	3. d	4. a	5. c
6. a	7. b	8. c	9. d	10. b
11. c	12. d	13. b	14. c	15. c
16. b	17. d	18. d	19. c	20. b
21. d	22. c	23. d	24. c	25. d
26. a	27. b	28. c	29. b	30. a
31. d	32. d	33. a	34. a	35. b
36. b	37. c	38. d	39. c	40. b
41. c	42. a	43. d	44. b	45. c
46. d	47. b	48. c	49. a	50. b
51. d	52. b	53. a	54. a	55. b
56. c	57. d	58. b	59. b	60. c
61. a				

PART IV

INSTRUCTOR'S GUIDE TO LABORATORY EXPERIMENTS FOR BASIC CHEMISTRY

Fifth Edition

EXPERIMENT 1

CLASSIFICATION

Time About 2 hours

Materials A selection of buttons (enough so that each group of students may have about 50); bar magnet; a selection of various chemical substances (labeled only by a letter code) in sealed test tubes or vials. Some substances that you may use for this part of the experiment are listed below.

Note: Seal all test tubes and vials.

Code	Material
A	Magnesium turnings
B	Aluminum turnings
C	Copper turnings
D	Iron turnings or powder
E	Zinc turnings or dust
F	Tin turnings
G	Nickel strips or small pieces of metal
H	Lead strips
I	Cadmium turnings or strips
J	Mercury (in a sealed vial for safety)
K	Bromine (in a sealed vial for safety)
L	Chlorine (in a sealed vial for safety)
M	Sulfur powder
N	Carbon (in the form of graphite)
O	Iodine crystals
P	Sodium chloride, $NaCl$
Q	Sodium bromide, $NaBr$
R	Cobalt(II) chloride, $CoCl_2$
S	Copper(II) sulfate, $CuSO_4 \cdot 5H_2O$
T	Cobalt(II) nitrate, $Co(NO_3) \cdot 6H_2O$
U	Iron(III) oxide, Fe_2O_3
V	Iron(II) oxide, FeO
W	Ammonium dichromate, $(NH_4)_2Cr_2O_7$
X	Iron(II) sulfide, FeS
Y	Sodium carbonate, Na_2CO_3

Code	Material
Z	Cobalt nitrate in solution (to give a red color), $Co(NO_3)_2 \cdot 6H_2O$
AA	Copper sulfate in solution (to give a blue color), $CuSO_4 \cdot 5H_2O$
BB	Cobalt chloride in solution (to give a pink color), $CoCl_2$
CC	Nickel nitrate in solution (to give a green color), $Ni(NO_3)_2 \cdot 6H_2O$
DD	Nickel chloride in solution (to give a green color), $NiCl_2 \cdot 6H_2O$

Before you dispense the materials, put them in the test tubes and then seal the test tubes. Label each test tube with the appropriate letter of the alphabet. If you do this, the students need only pick up their sets of test tubes and start to classify.

Note: If you have any fears about using bromine and chlorine in this type of situation, feel free to eliminate them from this experiment.

Discussion of the Experiment

We usually allow students to work in groups of two or three when they are doing this experiment. We hope that by letting them work in groups, we will encourage them to discuss among themselves the various ways to classify objects.

Some of the criteria used to classify buttons are size, color, shape, sex, number of holes, and so forth. Some of the criteria used to classify chemical substances are states of matter, color, texture, magnetism (or lack of it), and so forth. This experiment should serve as groundwork for Chapter 3 of the text, where there is a section on the classification of matter. In Experiment 3 of this laboratory manual, we shall again take up the classification of matter, using the system discussed in Chapter 3 of the text.

Some Possible Answers to the Questions

Because of the nature of the questions at the end of this experiment, we will not list any possible answers. However, in subsequent experiments, this section of the instructor's guide will contain possible answers to the questions.

EXPERIMENT 2

USE OF THE BALANCE: DETERMINING THE DENSITIES OF SOME COMMON OBJECTS

Time About 2 hours

Materials (For a section of 24 students working in groups of two): triple beam balances (12); aluminum weighing pans (12); graduated cylinders (12 of each size, 10 mL, 25 mL, 50 mL); metric rulers (12); objects of varying density (distilled water, small blocks of wood, metal cylinders, glass marbles, ethyl alcohol).

Discussion of the Experiment

From this experiment the students should learn:

1. How to use the triple-beam balance.

2. How to use other laboratory equipment so that they can gather data to determine the density of an object.

In this experiment the students should work in groups of two. This will enable them to check each other's readings on the balance. Each person should make individual observations or measurements, and then see if the observations or measurements agree.

It is very important that the instructor help the student to master the use of the balance. Most students have their greatest difficulty in learning how to read the vernier scale. If the triple-beam balance you are using has such a scale, show each student individually how to read the vernier scale. This is the practice we follow at Middlesex County College with our balances. (Our lab sections have about 25 students.) If the balance you are using has sliding weights, then a class demonstration is usually sufficient to teach the students how to weigh. It is also a good idea to check the students' first few weighings. In this manner you can make sure they are using the balance correctly.

The density experiment is designed to get students to use various measuring devices to obtain the density of some common objects. The procedure is written in outline form to guide the student through the experiment. The data tables require very specific information. This should help you in reading and grading the report sheets that the students hand in.

In this experiment we stress significant figures, as well as the precision of each measuring device. It is hoped that the student will express this precision in the data submitted. Too often our students have submitted calculated densities to six decimal places, and have seen nothing wrong in doing this.

The questions at the end of the experiment are designed to make the students examine the procedures they followed, discover any other possible procedures that could have been followed, and find some practical applications of this work.

Some Possible Answers to the Questions

(Please note that the following are suggested answers to the questions, based on our experience with this course over the years; however, they are not the only possible answers.)

1. The density would be less at 50°C than at 25°C, since alcohol expands when it is heated. Therefore a given mass of alcohol at 50°C occupies a greater volume than a similar mass at 25°C. Since the volume appears in the denominator of the formula for density, the greater the volume, the smaller the density.

2. The density of the human body is a little less than 1 g/cm^3. You could draw on your experiences with swimming or floating on water. Also in murder-mystery movies we're told that the dead body was found floating on the water. Since the density of water is 1 g/cm^3, anything that floats on water has a density less than 1 g/cm^3. You could measure the density of the human body in a bathroom by making use of a bathroom scale and the bathtub. Use the bathtub for finding your volume by water displacement, and the scale for finding your weight.

3. To calculate the weight of the big rock, first find the density of the small pieces that were chipped off the rock.

$$D = \frac{M}{V} = \frac{50.00 \text{ g}}{25.0 \text{ cm}^3} = 2.00 \text{ g/cm}^3$$

Now obtain the volume of the big rock. The volume of a cube is equal to its side cubed.

$$V = s^3 = (10.0 \text{ m})^3 = 1.00 \times 10^3 \text{ m}^3 = 1.00 \times 10^9 \text{ cm}^3$$

$$M = D \times V = (2.00 \text{ g/cm}^3)(1.00 \times 10^9 \text{ cm}^3) = 2.00 \times 10^9 \text{ g}$$

$$= 2.00 \times 10^6 \text{ kg}$$

$$= 4.40 \times 10^6 \text{ lb}$$

EXPERIMENT 3

SEPARATION OF SOLIDS FROM LIQUIDS

Time About 2 hours

Materials (For a section of 24 students working in groups of two): 1.0M lead(II) nitrate solution (1 liter); 0.20M potassium iodide solution (1 liter); 0.25M barium chloride solution (1 liter); 3M sulfuric acid (1 liter); 0.20M aluminum nitrate solution (1 liter); 6M ammonium hydroxide (1 liter); Buchner funnels (12); filter paper to fit Buchner funnels (1 box); suction flasks (12).

Discussion of the Experiment

The main purpose of this experiment is to introduce the student to the various methods of separating a solid from a liquid. The students are presented with three very different types of solids. Thus they can see both the advantages and the disadvantages of each method of separation for a particular type of solid.

Because we are not primarily concerned with chemical reactions in this experiment, it's good if you demonstrate the formation of each of the precipitates yourself. You can use sufficient quantities of reagents to produce enough products for distribution to the entire class. Then the student can concentrate on assembling the apparatus and performing the separations. (If you have a very large lab section, you may want the students to prepare the precipitates themselves.) We also recommend that you show the students how to fold the filter paper, how to arrange the suction and filtration apparatus, and how to work the centrifuge.

Caution the students that if they spill any of the materials on themselves, they should wash immediately.

Some Possible Answers to the Questions

(Please note that the following are suggested answers to the questions, based on our experience with this course over the years; however, they are not the only possible answers.)

1. Centrifuging
2. Decantation
3. Suction filtration and centrifuging

4. Decantation
5. Lead iodide and barium sulfate
6. Lead iodide and barium sulfate
7. None (all can be separated by centrifuging)
8. Barium sulfate and aluminum hydroxide

EXPERIMENT 4

USE OF THE GAS BURNER; STUDY OF ELEMENTS, COMPOUNDS, AND MIXTURES

Time About 2 hours

Materials (For a section of 24 students working in groups of two): Gas burners (12); pieces of cardboard (12); crucibles (12); crucible tongs (12); bar magnets (12); a selection of various substances in test tubes labeled by letter code. (Each group should have a set of test tubes with about 1 g of solid or 1 mL of liquid in each tube.) The following substances should be used.

Test tube	Contents
A	Mercury(II) oxide
B	Salt-sand mixture (50 percent salt, 50 percent sand)
C	Platinum wire attached to a glass rod
D	Cobalt(II) chloride dissolved in water (20 g of cobalt(II) chloride per 100 cm^3 of solution)
E	Iron and sulfur mixture (64 percent iron, 36 percent sulfur)

For ease of dispensing, put materials in letter-coded test tubes. If you use this system, the students need only pick up their sets of test tubes and start to classify.

Discussion of the Experiment

From this experiment the student should learn:

1. To recognize the parts of the laboratory gas burner, so he or she can use it properly.

2. To understand the differences between an element, compound, mixture, and solution by performing some simple tests.

In the first part of the experiment, each student should work on his or her own gas burner. Initially many students have trouble lighting the burner. This difficulty usually stems from two sources:

1. They don't know how to turn on the gas supply valve properly.

2. They don't know how to use the striker properly.

You might find it advantageous to demonstrate the use of the burner after the students have read the Introduction to the experiment, but before they actually try to use the burner.

Note: Be sure to emphasize the procedure to follow in case of a flashback.

In the second part of the experiment, the students should work in groups of two. This serves much the same purpose as in Experiment 1. It encourages students to interact with each other as they make decisions about what they observe. The students are exposed to two mixtures, a compound, a solution, and an element. At the end of the experiment, the students also form a compound from a mixture of iron and sulfur. In trying to classify these substances, the students get acquainted with various techniques used in the chemistry laboratory; for example, filtering and evaporating a liquid.

It is best to perform the decomposition of the mercury(II) oxide and the combination of iron and sulfur in a fume hood. Doing this enables students to avoid exposure to mercury vapor and sulfur dioxide fumes.

At the end of the experiment, it's a good idea to have a discussion of the limitations imposed on the students in testing each substance. You might also mention how hard it is to determine whether something is an element or a compound. Part of the discussion might involve the questions at the end of the experiment. Questions 2 and 4 are very difficult for students at this level to answer. However, some of the more interested students will go beyond this laboratory exercise in an attempt to answer these questions.

Some Possible Answers to the Questions

(Please note that the following are suggested answers to the questions, based on our experience with this course over the years; however, they are not the only possible answers.)

1. No, the match should not ignite, since the temperature in that part of the flame is not high enough to ignite the match.

2. To separate the alcohol-and-water solution, you might try distillation and take advantage of the difference in boiling points of the two substances.
3. You might try passing an electric current through the water to see whether it decomposes into its elements.
4. One test you might try is electrolysis of the molten solid, to break it down into its elements. You might weigh a small piece of the substance, then heat it. See if the substance gains or loses weight when it is heated. You might analyze the product obtained after heating to see whether it acts like the original substance.

EXPERIMENT 5

EMPIRICAL FORMULA OF A COMPOUND

Time About 2 hours

Materials (For a section of 24 students working in groups of two); magnesium ribbon (1 pkg); crucibles plus covers (24); crucible tongs (12); gas burners (24); balances (12).

Discussion of the Experiment

From this experiment the student should learn:

1. To perform an empirical-formula determination.
2. To tell the difference between an empirical and a molecular formula.
3. To use raw laboratory data to determine the empirical formula of magnesium oxide.

Supply each student with a strip of magnesium ribbon approximately 100 cm in length. (A 100-cm length of magnesium ribbon weighs about one gram.) It is best to vary the lengths so that each student has a different weight of magnesium. This shows students that varying the amount of magnesium has no effect on the results of the experiment. Use lengths of magnesium ribbon that are between 80 cm and 130 cm.

In order to save time and minimize error, use new or very clean crucibles. We use this tactic to eliminate the need to heat the crucibles to constant weight. If it is out of the question to use clean crucibles, then you should show your students how to heat their crucibles to constant weight.

Let the students work in groups of two, but see that each member of the group performs his or her own experiment. Each group will then have two sets of data for comparison.

It is very important that students coil the magnesium ribbon properly. The coil must rest on the bottom of the crucible so that it is heated to the fullest extent possible. Students should adjust the flame of the gas burner so that they obtain the hottest possible flame. Be sure that the crucible sits over this hot flame. One of the chief sources of error in this experiment is incomplete combustion. Another source of error in this experiment is using dirty magnesium ribbon. You may want to have your students clean the magnesium ribbon with some sandpaper to remove any magnesium oxide coating.

Keep students aware of the fact that they should not touch the hot crucible with their hands or with a paper towel, but that they should use crucible tongs. We recommend that you demonstrate the proper use of the crucible tongs to your students.

Some Possible Answers to the Questions

(Please note that the following are suggested answers to the questions, based on our experience with this course over the years; however, they are not the only possible answers.)

1. Yes, the empirical formula of the compound will be the same. The amount of magnesium you start with is unimportant.

2. If you had omitted Steps 12 and 13, the crucible would have contained a mixture of magnesium oxide (MgO) plus magnesium nitride (Mg_3N_2). A given weight of magnesium will produce a smaller weight of Mg_3N_2 than of MgO. Therefore the final weight of the crucible plus contents would have been lower than expected. To determine the amount of oxygen that combined with the magnesium, subtract the weight of the crucible plus cover plus magnesium from the final weight of the crucible plus cover plus contents after heating. This number would be much lower than expected. You would have a low oxygen reading. This, in turn, would cause the empirical formula of the compound to be $Mg_{>1}O_1$.

3. If the chemist had 50.00 g of S to start with and 100.00 g of sulfur-oxygen compound after heating, it would mean that 50.00 g of oxygen had combined with the sulfur.

$$\text{Moles of S atoms} = (50.00\ \cancel{g})\left(\frac{1.00\ \text{mole}}{32.0\ \cancel{g}}\right) = 1.56\ \text{moles}$$

$$\text{Moles of O atoms} = (50.00\ \cancel{g})\left(\frac{1.00\ \text{mole}}{16.0\ \cancel{g}}\right) = 3.12\ \text{moles}$$

The empirical formula of the compound is $S_{1.56}O_{3.12}$ or SO_2.

EXPERIMENT 6

ANALYZING UNSEEN THINGS

Time About 2 hours

Materials (For a section of 24 students working in groups of two): assorted boxes with various compartmental arrangements in each box, and different objects in each box (12 boxes); magnets (6); nichrome wire attached to a glass rod (12); 6M HCl (100 mL); sodium chloride (100 g); calcium chloride (100 g); potassium chloride (100 g); barium chloride (100 g); strontium chloride (100 g); gas burners (12); cobalt-blue glass plates (12); lithium chloride (100 g).

Discussion of the Experiment

From this experiment the student should learn:

1. How to use the scientific method to design a model.
2. How to perform flame tests.

For the first part of the experiment, prepare sealed boxes containing various objects. You may use small gift boxes or cigar boxes. Partition the interior of the boxes into halves, or thirds, or quarters. The easiest way to partition them is with cardboard and scotch tape, although you may also use wood to make partitions. Put in each box several objects, such as marbles, iron balls, styrofoam balls, and metal washers. We recommend that you place only one type of object in each box.

Prepare many boxes with numerous combinations in the placement of objects. For example, put together a series of boxes containing only iron balls. Let some of the boxes have one ball in each partitioned area. Let other boxes have two balls in one partitioned area, and one ball in each of the other areas, and so forth.

Code the boxes so that you are aware of what is inside. If the experiment is to be of value, it will be important for you to lead students in a meaningful inquiry. Each group should study at least two boxes.

In the second part of the experiment, students should have few problems in determining the flame colors of the known samples. The potassium compound is the only sample about which some question might arise. Remind students that they must use the cobalt-blue glass in order to eliminate the interference from sodium, and an interfering color from potassium.

After students have performed the test on the known samples, give them an unknown sample.

Some Possible Answers to the Questions

(Please note that the following are suggested answers to the questions, based on our experience with this course over the years; however, they are not the only possible answers.)

1. There are so many possible answers to this question that we shall not enumerate them here. Our students really come up with very interesting examples!

2. The electron configurations in each of the metal ions are different and so are the numbers of electrons; therefore different amounts of energy are involved in the different transitions that occur, and different colors of light are produced.

EXPERIMENT 7

ODOR SENSITIVITY

Time About one hour to perform the experiment and one hour to discuss the results

Materials (For a section of 24 students working in groups of two): thirty 4-ounce bottles with covers; vanillin (or a bottle of vanilla extract); isopentyl acetate (banana oil) (an alternative for isopentyl acetate is butyl acetate); butyric acid (rotted cheese odor).

PREPARATION OF SOLUTIONS

1. Vanillin

Stock solution (0.4 percent-by-weight)
Dissolve 1 g of vanillin in 250 mL of distilled water; warm if necessary.

Solution Van-1 (0.1 percent)
Mix 25 mL of stock solution with enough distilled water to yield 100 mL of solution.

Solution Van-2 (0.01 percent)
Mix 10 mL of stock solution with enough distilled water to yield 400 mL of solution.

Solution Van-3 (0.001 percent)
Mix 1 mL of stock solution with enough distilled water to yield 400 mL of solution.

2. Isopentyl acetate (or butyl acetate)

Solution Ban-1 (0.1 percent)
Dissolve 0.1 g of isopentyl acetate in enough distilled water to yield 100 mL of solution.

Solution Ban-2 (0.01 percent)
Dilute 10 mL of solution Ban-1 with enough distilled water to yield 100 mL of solution.

Solution Ban-3 (0.001 percent)
Dilute 1 mL of solution Ban-1 with enough distilled water to yield 100 mL of solution.

3. Butyric acid (Prepare under the hood.)

Stock solution (0.1 percent)
Dissolve 0.1 g of butyric acid in enough distilled water to yield 100 mL of solution.

Solution Che-1 (0.01 percent)
Dilute 10 mL of stock solution with enough distilled water to yield 100 mL of solution.

Solution Che-2 (0.001 percent)
Dilute 1 mL of stock solution with enough distilled water to yield 100 mL of solution.

Solution Che-3 (0.0001 percent)
Dilute 1 mL of stock solution with enough distilled water to yield 1000 mL of solution.

DIRECTIONS FOR FILLING THE THIRTY BOTTLES

1. Divide the 30 bottles into 10 groups of three sets of bottles.
2. Number the bottles as shown below.

1-A	1-B	1-C
2-A	2-B	2-C
⋮	⋮	⋮
10-A	10-B	10-C

3. Using the following key, fill each bottle about half full with the appropriate sample.

Bottle set	Bottle designation		
	A	B	C
1	Water	Van-1	Water
2	Water	Water	Che-1
3	Water	Water	Ban-1
4	Water	Water	Water
5	Che-2	Water	Water
6	Ban-2	Water	Water
7	Water	Water	Van-2
8	Ban-3	Water	Water
9	Water	Che-3	Water
10	Water	Van-3	Water

Discussion of the Experiment

Ask the students to work in groups of two. Stress that each group should perform its work independently of other groups. After students have finished the experiment, the groups will be able to compare their results during the discussion session.

Note: If you have students with existing respiratory diseases, you may not want them to perform this experiment.

One of the problems that students will have with this experiment is odor fatigue. You can overcome this problem by having students take short rest periods between inhalations of each sample. Make sure that caps as well as the bottles are numbered; otherwise there will be contamination of the nonodorous substances if the students replace the caps incorrectly. Also be sure that the samples are fresh (no older than two weeks). Odor deteriorates as the sample ages. Distilled water tends to develop a stagnant odor after a period of time.

Students sometimes try to see whether a bottle contains an odorous substance by looking for some color in the sample. Ask them not to do this. Students are to use their sense of smell, not sight. In recent years we have solved this problem by adding food coloring to all the bottles. It makes for a very appealing display!

Students will also ask how they can tell whether an odor deserves a rating of 4 or 5. This is a problem they will have to solve for themselves. The differences between a strong odor and a very strong odor are highly subjective.

The object of this experiment is to encourage the student to make decisions and use his or her sense of smell. In the process, the student will learn a standard procedure for evaluating odors.

Some Possible Answers to the Questions

(Please note that the following are suggested answers to the questions, based on our experience with this course over the years; however, they are not the only possible answers.)

1. The major problem students have will probably be odor fatigue.

2. Some other ways of quantifying odor are the odor-intensity index (OII) and the threshold odor concentration (TOC). The odor-intensity index is the number of times the original sample is diluted by half with odor-free water before one obtains odor-free water. The threshold odor concentration is the concentration of a substance (in milligrams per liter) that renders the odor just barely detectable.

LABORATORY EXERCISE I

ARRANGING TWENTY-ONE ELEMENTS

Time About one hour to perform, plus one hour for discussion

Materials 21 cards containing information about 21 elements

Discussion of the Exercise

This is a simple exercise, in which the student will be able to discover for himself/herself how elements can be arranged in logical order according to their characteristics. The information on the cards generally corresponds to the first 21 elements in the periodic table. It is important that the student does not take these cards and copy the arrangement of the modern-day periodic table. He/she must rely on his/her own ideas to arrive at an arrangement. It is also important that the student realize that there is no correct answer, since this is basically a thought exercise. The student can arrange and rearrange the cards as he/she sees fit.

This exercise has been used very successfully in class testing. Students worked alone and in groups to arrange the cards. Some students came up with triads, and others arranged the elements in octaves, similar to the historical treatments of the elements. Some of the brighter students came close to the arrangement of the present-day table.

This is a very useful exercise to refer to when you are discussing the history of the periodic table. The instructor should use this as an introduction to the periodic table. The timing is important, because once you have discussed the arrangement of the periodic table, the value of this exercise is lost. After hearing about Mendeleev's discoveries, students will no longer be inclined to discover their own arrangements; most probably, they will copy Mendeleev's.

Summary of the Information on the Cards in the Laboratory Manual

Atomic weight	No. of hydrogens in hydride	No. of fluorines in fluoride	No. of oxygens in oxide	Ionization potential (kcal/mole)
1	1	1	0.5	314
4	-	-	-	567
7	1	1	0.5	124
9	2	2	1	215
10	3	3	1.5	190
12	4	4	2	260
14	3	3	2.5	335
16	2	2	1	314
19	1	1	0.5	402
20	-	-	-	497
23	1	1	0.5	119
24	2	2	1	176
27	3	3	1.5	138
28	4	4	2	188
31	3	3	2.5	254
32	2	2	3	239
36	1	1	0.5	300
40	-	-	-	363
39	1	1	0.5	100
40	2	2	1	141
45	-	3	1.5	151

EXPERIMENT 8

THE PERIODIC TABLE: THE CHEMISTRY OF ELEMENTS WITHIN A GROUP

Time About 2 hours

Materials (For a section of 24 students working in groups of two: lithium chloride, sodium chloride, and potassium chloride; carbon chips, silicon chips, and germanium chips; tin strips and lead strips (a few crystals of each of the chips in sealed test tubes and small pieces of the tin and lead in test tubes). (Have a set of test tubes for each group of students if possible.) Strontium oxide (1 bottle); barium oxide (1 bottle); copper turnings (1 bottle); 6M nitric acid (100 mL); phosphorus pentoxide (1 bottle); arsenic pentoxide (1 bottle); bar magnets (12); neutral litmus paper (1 pkg).

Discussion of the Experiment

This experiment enables the student to examine some of the properties of elements within a group. We are interested in having students observe the similarities that exist between elements within a group. To ensure safety, one must observe certain precautions, which we shall discuss in the following paragraphs. We realize that some instructors may feel that some of the chemicals used in this experiment are dangerous. However, we would like to point out that the quantities used by the student are extremely small and present no real hazard.

PART 1: THE GROUP IVA ELEMENTS

This part of the experiment presents no special difficulties. It's better if the lead metal is in the form of strips rather than dust. Caution students to wash their hands if they touch the elements.

PART 2: THE GROUP IIA METAL OXIDES

We suggest that you dispense each of the materials (SrO, BaO, and CaO) yourself for this part of the experiment. Repeatedly caution the students not to inhale or touch the materials either before or after they are placed into solution. Warn the students that if any of the chemicals come into contact with their skin, they should wash the skin immediately with plenty of soap and water.

PART 3: THE GROUP VA NONMETAL OXIDES

This part of the experiment requires extra-special care in handling the materials. You may want to do this as a demonstration. You should dispense the phosphorus pentoxide yourself, cautioning students not to touch or inhale the materials.

When students prepare nitrogen dioxide (NO_2), they should do so under a fume hood. They will prepare the NO_2 using copper and 6M HNO_3.

$$Cu + 4HNO_3 \longrightarrow Cu(NO_3)_2 + 2H_2O + 2NO_2$$

Again, caution students not to breathe the gas.

When it is dissolved in water, the nitrogen dioxide will form nitrous and nitric acids. To simplify the procedure, ask the students to wet a piece of neutral litmus paper with distilled water, then to hold it over the test tube generating the NO_2 gas. The neutral litmus paper should turn red.

PART 4: THE GROUP IA METALS (A Demonstration)

Perform this experiment under a fume hood. Each of the materials involved (Li, Na, and K) is usually stored under oil. Obtain small pieces of each material and place them in a beaker containing dry pentane. The pentane will dissolve the oil. When performing the experiment, use extra caution, since each of these materials reacts violently with water.

Since potassium is the most reactive of the three metals, it is likely to ignite the hydrogen produced in the reaction with water, so be prepared for this.

To dispose of the extra materials safely, react any excess of the materials with methanol. Also wash the beakers containing the pentane with methanol, since small bits of each metal might remain in the beakers.

Caution: Use only a small piece, about the size of a pea, of each metal. Drop each piece of metal into a separate beaker (250-mL capacity) half filled with water.

Some Possible Answers to Questions

(Please note that the following are sugggested answers to the questions, based on our experience with this course over the years; however, they are not the only possible answers.)

1. Elements within a group have similar chemical properties because they have similar numbers of electrons in their outermost energy levels.
2. The differences in the speed of reaction for Li, Na, and K is due in part to the positions of their outermost electrons.

	Li	Na	K
Outermost electron	$2s^1$	$3s^1$	$4s^1$

The reactions performed involved the removal of the outermost electron from each of the metals. The farther away the outermost electron is from the nucleus, the easier and faster it can be removed. The potassium was the fastest to react, then the sodium, and last the lithium.

3, 4, 5. These are discussion questions; therefore they have many possible answers.

EXPERIMENT 9

THE LAW OF DEFINITE COMPOSITION

Time About 2 hours

Materials (For a section of 24 students): test tubes (heavy-walled, clean, and dry, 24); potassium chlorate (100 g).

Discussion of the Experiment

Before the students perform this experiment, introduce the class to the law of definite composition in your lecture sessions or your assigned textbook readings. Unless you do this, students may not have much understanding or appreciation for what they are about to do.

The procedure is quite simple, so the student shouldn't run into much difficulty in carrying it out. Be sure to caution the student to heat the potassium chlorate gently at first, then with greater amounts of heat as the material starts to decompose.

Some instructors have the student test for oxygen gas by using a glowing splint, as a check to see whether all the potassium chlorate has been decomposed.

Caution: Potassium chlorate is a strong oxidizing agent. A mixture of potassium chlorate and various organic chemicals can lead to a fire. Be sure that no organic chemicals come into contact with the heated potassium chlorate.

Some Possible Answers to the Questions

(Please note that the following is a suggested answer to Question 2, based on our experience with this course over the years; however, it is not the only possible answer.)

2. There are two ways to check whether all the potassium chlorate has been decomposed.
   (a) Use of a glowing splint to see if oxygen gas is still being liberated
   (b) Heating the test tube and contents repeatedly, weighing after each heating, until constant weight is reached

EXPERIMENT 10

TYPES OF CHEMICAL REACTIONS

Time About 2 hours

Materials (For a section of 24 students working in groups of two): iron dust or powder (200 g); sulfur powder (100 g); mercury(II) oxide (25 g); mossy zinc (100 g); 6M hydrochloric acid (100 mL); 0.5M lead(II) nitrate (500 mL); 1M potassium iodide (500 mL).

Discussion of the Experiment

This experiment enables students:

1. To perform various chemical reactions that they have read about in their textbooks.

2. To use various laboratory techniques that they have been learning in their previous laboratory work.

The reactions performed in this experiment create no special difficulties for the student. However, students must familiarize themselves with each procedure before they attempt the reaction.

PART 1: COMBINATION REACTIONS

Students should heat the iron and sulfur under a hood. This will minimize any noxious fumes that the burning sulfur may create.

PART 2: DECOMPOSITION REACTIONS

Caution the students not to inhale or touch the mercury that forms in the test tube as a result of the decomposition of the mercury(II) oxide. Also tell them how to dispose of the contents of the test tube properly.

PART 3: SINGLE-REPLACEMENT REACTIONS

This reaction presents no special problems. However, tell students to be ready to collect and test for the presence of hydrogen immediately, as soon as the reaction is completed.

PART 4: DOUBLE-REPLACEMENT REACTIONS

When the student is separating the lead iodide from the solution, the filter paper may become clogged. Explain to the students that they do not need to collect every last drop of filtrate before evaporating it.

Some Possible Answers to the Questions

(Please note that the following are suggested answers to the questions, based on our experience with this course over the years; however, they are not the only possible answers.)

1. (a) Single-replacement (b) Double-replacement
   (c) Decomposition (d) Combination (e) Combination

2. Hydrogen chloride fumes are produced as the filtrate is evaporated.

EXPERIMENT 11

DETERMINATION OF THE AMOUNT OF PHOSPHATE IN WATER

Time About 2 hours

Materials (For a section of 24 students working in groups of two): colorimeters (12 if possible or fewer will do); standard phosphate solutions (directions for preparation follow); color reagent (prepare one liter of this solution if each group does one unknown and four standards). Note that the directions to follow (number 5) give amounts to prepare 100 mL of color reagent. Therefore you will have to multiply each quantity by ten. Unknown phosphate solutions (same as knowns); 100 mL beakers (60); 25-mL graduated cylinders (12); 5-mL graduated cylinders (12).

PREPARATION OF REAGENTS

1. Sulfuric acid (5N)
   Dissolve 70 mL of concentrated sulfuric acid in enough water to make 500 mL of solution.

2. Potassium antimonyl tartrate solution
   Weigh 0.3 g of $K(SbO)C_4H_4O_6 \cdot (1/2) H_2O$. Dissolve this in 50 mL of water, then pour it into a 100-mL volumetric flask and dilute to mark.

3. Ammonium molybdate solution
   Prepare by dissolving 4 g of $(NH_4)_6Mo_7O_{24} \cdot 4H_2O$ in 100 mL of distilled water. Store in a plastic bottle at 4°C.

4. Ascorbic acid solution (0.1M)
   Dissolve 1.8 g of ascorbic acid in 100 mL of distilled water. The solution is stable for about a week if stored at 4°C.

5. Combined reagent (also called color reagent).
   Mix the reagents listed above in the following proportions for 100 mL of the mixed reagent:
   50 mL of 5N sulfuric acid
   5 mL of potassium antimonyl tartrate solution
   15 mL of ammonium molybdate solution
   30 mL of ascorbic acid solution

Mix after addition of each reagent. All reagents must reach room temperature before they are mixed and must be mixed in the order given. If the combined reagent becomes turbid, shake the solution and let it stand for a few minutes until the turbidity disappears. The stability of this reagent is limited; it must be prepared fresh each day.

6. Stock phosphate solution
   Dissolve 0.4393 g of predried KH_2PO_4 (potassium dihydrogen phosphate) in distilled water and dilute to 1 liter. This solution is such that 1 mL = 0.1 mg P.

7. Standard phosphate solution A
   Dilute 100 mL of stock solution to 1 liter with distilled water. This solution is such that 1 mL = 0.01 mg P.

8. Standard phosphate solution B
   Dilute 100 mL of solution A to 1 liter with distilled water. This solution is such that 1 mL = 0.001 mg P.

PREPARATION OF STANDARDS FOR COLORIMETRIC DETERMINATION

Prepare a series of standards by diluting suitable volumes of solutions A and B to 100 mL with distilled water. The following dilutions are suggested.

Milliliters of solution B	Concentration of solution in ppm of P
2.0	0.02
5.0	0.05
10.0	0.10

Milliliters of solution A	Concentration of solution in ppm of P
2.0	0.20
5.0	0.50
8.0	0.80
10.0	1.00

You may also use these solutions as the unknown solutions that you will give to the students.

Prepare all solutions, including the color reagent, prior to the laboratory session so that they are ready for student use.

Discussion of the Experiment

The main purpose of this experiment is to give the student an opportunity to work with a laboratory instrument. We chose the determination of phosphorus in order to increase the student's interest in the experiment. Depending on the ability of the student, you may or may not want to discuss the chemistry of the determination. Personally, we feel this is not necessary, since it is not one of the objectives of the experiment. We feel it is more important to spend time on the basic principle behind a colorimetric determination.

Be sure to point out to the student the proper placement of the colorimeter test tube in the sample compartment; tell students to close the sample compartment cover before taking a reading. The Spectronic-20 tubes are etched and should be placed in the sample compartment with the etching aligned with the mark on the sample holder.

You may want to have your students measure their samples with a pipette instead of with a graduated cylinder. If so, you should have a supply of 25-mL pipettes for the samples and a supply of 5-mL pipettes for the color reagent.

In the pre-laboratory discussion of the experiment, explain the ppm concentration unit. In the post-laboratory discussion you may want to explain graphing techniques.

Some Possible Answers to the Questions

(Please note that the following are suggested answers to the questions, based on our experience with this course over the years; however, they are not the only possible answers.)

1. The type of phosphate usually found in detergents is called polyphosphate. A typical polyphosphate is $Na_5P_3O_{10}$, sodium tripolyphosphate. The $(P_3O_{10})^{-5}$ is an example of a polyphosphate ion.

2. The chief sources of phosphates in the nation's waterways, lakes, and water supplies are detergents, agricultural fertilizer runoff, human wastes.

3. Dilute the original unknown solution with distilled water. For example, you might try diluting 25 mL of unknown solution with 25 mL of water. This cuts the concentration of the original solution in half. Now use 25 mL of this diluted solution with 5 mL of color reagent and see if you obtain a reading on the colorimeter. If the absorbance of the solution is measurable, the problem is solved. Remember, however, to multiply the concentration of phosphate found in the diluted solution by two, since the concentration of the phosphate in this solution is one-half that of the original solution. If the color of the solution is still too dark, then further dilutions are necessary.

EXPERIMENT 12

MEASURING THE pH OF SOME ACIDS, BASES, AND SALTS

Time About 2 hours

Materials (For a section of 24 students working in groups of two): pH meters (12 or as many as possible); pH paper or universal indicator (1 pkg of paper or 500 mL of universal indicator); 1M ammonium chloride (50 mL); 1M potassium chloride (50 mL); 1M sodium carbonate (50 mL); 0.1M sodium hydroxide (50 mL); 0.1M hydrochloric acid (50 mL); 0.1M ammonium hydroxide (50 mL); vinegar or 4 percent-by-weight acetic acid solution (1 liter); orange juice (1 liter); milk (1 liter); a soft drink (1 liter); buffer solution for standardization of pH meter (1 liter).

Discussion of the Experiment

The main objective of this experiment is to have the student learn how to perform pH measurements. Students will also learn about the pH scale and the relative strengths of acids and bases. Included in the items whose pH's will be determined are some very common substances (for example, orange juice and vinegar). We hope that this will give the student a better grasp of the concept of pH.

The experiment is divided into two parts:

1. Determination of pH with pH paper (or universal indicator)
2. Determination of pH with a pH meter

The student should not encounter many problems in Part 1 of the experiment. If you give students universal indicator to use, be sure to tell them the quantity of indicator they need for a 2-mL sample. The student may have a hard time determining the pH of the orange juice using universal indicator. This is fine, since it will lead the student to think about other ways to determine the pH of colored materials.

If there aren't enough pH meters available for the students to use, you may do Part 2 of the experiment yourself as a demonstration. If students are to perform Part 2, be sure to give them detailed instructions on the use of the pH meter.

After completing both parts of the experiment, the student ought to be able to see the advantages and disadvantages of the two methods. The questions at the end of the experiment ask for precisely this information.

Although we have purposely stayed away from any mathematical discussion of pH, Question 3 at the end of the experiment brings up an important point that you might want to discuss more fully.

Some Possible Answers to the Questions

(Please note that the following is a suggested answer to **Question 3**, based on our experience with this course over the years; however, it is not the only possible answer.)

3. A difference of 2 pH units equals a magnitude of 10×10. In other words, the substance whose pH = 3 is 100 times as acidic as the substance whose pH = 5.

EXPERIMENT 13

WATER OF HYDRATION: THE FORMULA OF A HYDRATE

Time About 2 hours

Materials (For a section of 24 students); various hydrated salts listed below for unknowns (about 100 g of each); other hydrated salts may be added or substituted for the ones on this list. Gas burners (24); crucibles, covers, and crucible tongs (24 of each).

Compound	MW of anhydrous salt
1. $Na_2CO_3 \cdot 10H_2O$	106
2. $CuSO_4 \cdot 5H_2O$	160
3. $NiSO_4 \cdot 7H_2O$	155
4. $BaCl_2 \cdot 2H_2O$	208
5. $CaSO_4 \cdot 2H_2O$	136
6. $KAl(SO_4)_2 \cdot 12H_2O$	258

Discussion of the Experiment

The major objective of this experiment is to enable the student to examine the relationship between salts and their waters of hydration. The experiment is direct and simple. However, the calculations require the student to have a working knowledge of ratio and moles. These are important concepts that the student should use.

The student should use between 2 g and 3 g of hydrated salt. The sample should be finely ground in order to ensure a large surface area, so that the water can be driven off. Tell the students to heat their samples over the hottest part of the burner flame, after an initial period of gentle heating.

Some Possible Answers to the Questions

(Please note that the following are suggested answers to the questions, based on our experience with this course over the years; however, they are not the only possible answers.)

1. An anhydrous compound, possibly $CuSO_4$, is picking up water, and becoming $CuSO_4 \cdot 5H_2O$.

2. Some possible reasons for a student's inability to get his or her hydrate to reach constant weight are:
   (a) The sample was not ground finely enough and is physically entrapping water.
   (b) High humidity in the laboratory is causing the sample to pick up moisture upon cooling.
   (c) The student is using different balances that are not properly zeroed.

EXPERIMENT 14

THE ACETIC ACID CONTENT OF VINEGAR

Time About 2 hours

Materials (For a section of 24 students): commercial colorless vinegars (2 liters); Erlenmeyer flasks, 125-mL size (24); phenolphthalein indicator solution (100 mL); 1.25M standard sodium hydroxide solution (2 liters) (this solution contains 0.050 g of NaOH per milliliter of solution); burettes (24).

Discussion of the Experiment

The major objective of this experiment is to have students perform an acid-base titration. Before allowing students to proceed with the experiment, you should demonstrate the proper technique in the use of the burette and perform a sample analysis so that the students can see the entire procedure from start to finish. Be sure to show the students how to read the burette properly and emphasize the fact that there should be no air bubbles in the tip of the burette.

You may want to review the theory behind acid-base titrations and go through some sample calculations. However, the major problem for students seems to be overshooting the endpoint. Thus it's more important to concentrate on proper technique in handling and reading the burette.

Some Possible Answers to the Questions

(Please note that the following are suggested answers to the questions, based on our experience with this course over the years; however, they are not the only possible answers.)

1. Any acid present in the vinegar sample will react with the NaOH solution and will be recorded as acetic acid. Therefore your results will show a greater percentage of acetic acid than was actually in the sample.

2. To avoid dilution or contamination of your titrating reagent, you should always rinse your burette with a small amount of the solution you're going to put in it.

EXPERIMENT 15

WATER HARDNESS: A TITRATION ANALYSIS

Time About 2 hours

Materials (For a section of 24 students): burettes (24); EDTA-Mg buffer solution (250 mL, prepared according to the directions to follow); Eriochrome black T indicator mixture (100 g, prepared according to the directions to follow); 0.01M EDTA solution (1 liter, prepared according to the directions to follow); 250-mL Erlenmeyer flasks (24); 25-mL pipettes (24) ; 25-mL graduated cylinders (24) ; 100-mL beakers (24) ; hardness solutions (prepared according to the directions to follow).

PREPARATION OF REAGENTS

1. Buffer solution
   Dissolve 16.9 g of NH_4Cl in 143 mL of concentrated NH_4OH. Add 1.25 g of the magnesium salt of EDTA and dilute to 250 mL with distilled water. Keep tightly closed in a plastic container. Discard as soon as 1 or 2 mL fails to produce a pH of 10 at the endpoint of the titration.

2. Standard 0.01M EDTA
   Weigh 3.723 g of dry reagent-grade sodium salt of EDTA into a 1-liter volumetric flask and dilute with water to the mark.

3. Eriochrome black T
   Using a mortar and pestle, mix 0.5 g of Eriochrome black T and 100 g of NaCl.

Discussion of the Experiment

The main objective of this experiment is for students to learn how to determine the hardness of water using a volumetric analysis technique. Students will also learn how to use some volumetric equipment like burettes and pipettes.

Before allowing students to proceed with the experiment, you should demonstrate the proper use of the burette and the pipette and perform a sample analysis, so that students can see the entire operation from start to finish.

To add more relevance to the experiment, encourage students to bring in their own samples to analyze. If you wish, you can prepare solutions of standard hardness by using the following directions.

PREPARATION OF STANDARD HARDNESS SOLUTIONS

1. Preparation of stock solution of $CaCO_3$ (1000 ppm)

   Weigh exactly 1.000 g of analytical-reagent-grade $CaCO_3$ and dissolve it in a little dilute hydrochloric acid. Transfer this solution to a 1-liter volumetric flask and add water to the mark.

   Note: You might prefer to use $CaCl_2$ in place of $CaCO_3$ to make up the standard hardness solutions. Be sure to make the proper calculations if you do this.

2. Preparation of samples of known hardness

   Prepare a series of standard hardness solutions by diluting suitable volumes of stock solution to 100 mL with distilled water. The following dilutions are suggested.

Milliliters of $CaCO_3$ stock solution needed for preparation of standard	1	2	3	4	5	7	10
Hardness of the standard solution prepared, ppm of $CaCO_3$	10	20	30	40	50	70	100

Some Possible Answers to the Questions

(Please note that the following are suggested answers to the questions, based on our experience with this course over the years; however, they are not the only possible answers.)

1. If the pH is too high, the Ca^{+2} and Mg^{+2} ions will precipitate from the solution as $Ca(OH)_2$ and $Mg(OH)_2$.

2. Make quantitative dilutions of the original sample.

EXPERIMENT 16

CHARLES'S LAW: A LOOK AT ONE OF THE GAS LAWS

Time About 2 hours

Materials (For a section of 24 students working in groups of two): 125-mL Erlenmeyer flasks (12); thermometers (12); glass tubing (a total of 8 m in length, of diameter to fit two-hole rubber stoppers); 2-hole rubber stoppers (12, to fit Erlenmeyer flasks).

Discussion of the Experiment

This experiment enables students:

1. To set up an experimental apparatus to test Charles's law.
2. To check the validity of Charles's law.

Students must have completed the study of gases before they try this experiment. In contrast to the preceding experiments, this one may be difficult for the student to understand. Most of our basic chemistry students have a hard time understanding gases; they seem to find the topic too abstract. This experiment gives students a chance to generate their own data and test the validity of one of the gas laws. Most of the rigor in this experiment lies in the calculations. Students have their greatest difficulty in trying to understand how the water vapor affects the collection of a gas over water. This part of the calculation needs special attention. You may wish to review this with the students after the laboratory session.

Students shouldn't have much trouble with the experimental procedure. They might need some help in understanding why they must adjust the water level of the graduated cylinder to obtain the volume of the gas collected inside. If at all possible, try to check each student's apparatus for tightness of fittings, and so forth. (Instructors with large labs may not find this feasible.)

Some Possible Answers to the Questions

(The questions in this experiment are different from the type of question we have been asking in other experiments. We felt that the subject matter in the experiment is difficult enough without adding our usual type of question.)

1. Kelvin 2. Pressure 3. 10 liters

EXPERIMENT 17

MELTING POINTS AND BOILING POINTS OF SOME ORGANIC COMPOUNDS

Time About 2 hours

Materials (For a section of 24 students working in groups of two): mineral oil (2 liters); benzoic acid (100 g); citric acid (100 g); naphthalene (100 g); diphenyl (100 g); urea (100 g); ethyl alcohol (100 mL); ethylacetate (100 mL); isopropyl alcohol (100 mL); thermometers (12); melting-point capillary tubes (1 pkg).

Discussion of the Experiment

The main objective of this experiment is to enable the student to develop a technique for determining the melting points and boiling points of compounds.

Although students may feel that the experiment appears easy, there are many hidden obstacles that they may encounter. We shall review these obstacles so that you may point them out. One problem occurs in the determination of melting points. If the compounds are not finely ground, students may have a hard time filling the capillary tubes. You may need a mortar and pestle to grind the compounds. In addition, the rate of heating is extremely important in the determination of melting points if the student is to obtain reliable data. The rate of heating should not exceed 3°C per minute (even less in the area of the melting point of the compound). In the determination of boiling points, students may have difficulty in getting their solutions to boil gently. To ease this problem, you may want to introduce the use of boiling chips. Remind students that they cannot leave their gas burner under the test tube and expect to control the rate of heating. They must hold the burner and move it back and forth under the test tube until they achieve a desired rate of heating. They should practice with water until they have developed a good technique.

Melting Points of the Compounds

Diphenyl: 71°C
Naphthalene: 80°C
Benzoic acid: 122°C
Urea: 132°C
Citric acid: 153°C

Boiling Points of the Compounds (at 1 atm)

Ethylacetate: 77°C
Ethyl alcohol: 78°C
Isopropyl alcohol: 82°C
Water: 100°C

Some Possible Answers to the Questions

(Please note that the following are suggested answers to the questions, based on our experience with this course over the years; however, they are not the only possible answers.)

1. (a) An increase in atmospheric pressure increases the boiling point.
   (b) A decrease in atmospheric pressure decreases the boiling point.

2. You could perform a mixed melting point.

3. Once the liquid is completely vaporized, you would be heating air in the test tube. This would give erroneous results.

4. The melting points of some of the compounds you determined are greater than the boiling point of water.

EXPERIMENT 18

THE PERSONAL AIR POLLUTION TEST: THE ANALYSIS OF SOLIDS IN CIGARETTE SMOKE

Time About 2 hours

Materials (For a section of 24 students working in groups of two): cigarettes (filter tip, non-filter tip and "little cigars," 1 pkg of each); 250-mL suction flasks (12); filter paper (1 pkg); 1-hole rubber stoppers (12, to fit suction flasks); glass tubing (2 meters, with diameter to fit 1-hole rubber stoppers); rubber tubing (to fit over glass tubing on one end and cigarettes on the other); also rubber tubing for suction end of flask and aspirator).

Discussion of the Experiment

The two main objectives of this experiment are:

1. To inform the student of some of the dangers of smoking.
2. To enable the student to perform an analysis of solids in cigarette smoke so that he or she can compare the relative amounts of solids collected from different types of cigarettes.

The experiment is very straightforward and should not present any procedural difficulty to the student. The only variable the student has to work with is adjusting the aspirator so that the cigarette burns slowly.

If analytical balances are available, you may want to have the students use them for this experiment. It makes for a better experiment if students record all weighings to 0.1 mg.

Be sure that the laboratory has been cleared of all large containers of flammable materials before beginning this experiment.

We suggest that after the experiment has been performed, you hold a discussion period. The discussion can begin with the questions given at the end of the laboratory report.

Some Possible Answers to the Questions

(Please note that the following are suggested answers to the questions, based on our experience with this course over the years; however, they are not the only possible answers.)

1. Usually the filter-tip cigarette, with the filter removed, produces the most solids.

2. The final portion of the cigarette usually yields the most solids. By this time, the filter is working inefficiently.

EXPERIMENT 19

VOLTAIC AND ELECTROLYTIC CELLS

Time About 2 hours

Materials (For a section of 24 students working in groups of two : voltaic cell units*(12); zinc strips (12) for use as zinc electrodes; copper strips (12) for use as copper electrodes; 1M copper(II) sulfate (1 liter); saturated sodium chloride solution (1 liter); phenolphthalein indicator solution (100 mL); copper wire (thin wire works best) (1 spool); magnesium ribbon (1 pkg) : doorbells that operate on 1½ volts (12 or fewer if necessary); alligator clips (12 or fewer if you're using fewer doorbells; 3M sulfuric acid (1 liter); voltmeters (12, or as many as you can obtain).

Discussion of the Experiment

The main objectives of this experiment are for the student to:

1. Assemble a voltaic and an electrolytic cell.
2. See firsthand how to operate a voltaic cell by generating electricity from a chemical reaction.
3. See firsthand how to use electricity to produce a chemical reaction.
4. Learn the definition of a voltaic cell.
5. Learn the definition of an electrolytic cell.

In this experiment, we did not explain the theory of electro-chemical cells, since we believe that the topic is not appropriate for a basic chemistry course. In our opinion, the objectives stated above are the appropriate ones. If the student studies the diagrams of the cells carefully, he or she should have no difficulty in assembling the apparatus for the experiment.

---

* The voltaic cell units we use at our school were purchased from Morris and Lee (School Science Equipment), 1685 Elmwood Avenue, Buffalo, New York 14207.

An alternative way of performing the experiment is to use beakers instead of a voltaic cell unit. If you do this, you must construct a salt bridge out of glass tubing and explain its function to the students (Figure 1).

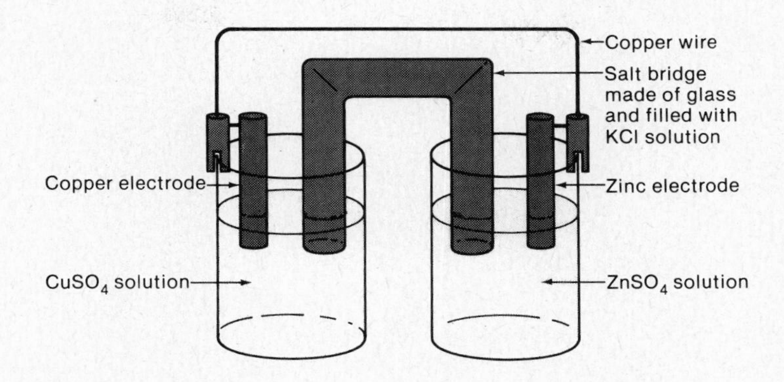

Figure 1 Using beakers to assemble a voltaic cell

Some Possible Answers to the Questions

(Please note that the following are suggested answers to the questions, based on our experience with this course over the years; however, they are not the only possible answers.)

1. Assemble a number of cells and connect them in series.

2. An automobile battery is a voltaic cell when it is being used to start the engine and an electrolytic cell when it is being recharged.

3. Many answers are possible.

EXPERIMENT 20

RADIOACTIVITY

Time About 2 hours

Materials (For a section of 24 students working in groups of two): Geiger counters (12); beta sources (12); alpha sources (12); gamma sources (12); lead strips, of uniform thickness and size (10 shields are needed by each group, but they can be passed around); meter sticks (12); calipers or centimeter rulers (12).

Discussion of the Experiment

The main objectives of this experiment are for the students to:

1. Operate a Geiger counter.

2. Observe the differences between the penetrating ability of beta radiation and the penetrating ability of gamma radiation.

We have been as specific as possible about the procedural aspects of this experiment. However, there are numerous types of Geiger counters, and each operates in a slightly different manner. Also, the degrees of radioactivity of samples are quite different, and you may want your students to make counts of longer or shorter duration than those suggested in the experiment.

PART 1: DETERMINING PLATEAU VOLTAGE

Have graph paper available so that students can plot their graphs immediately after obtaining their data. Caution students about exceeding the plateau range. It's best not to let them go beyond 1100 volts; otherwise they might burn out the tube. If your Geiger tubes are fairly new, you may want to have them stop at a lower voltage.

PART 2: DETERMINING BACKGROUND RADIATION

Make sure that students have removed all radioactive materials from their work areas before starting their background counts. If your counters read directly in counts per minute, have students take a series of background readings and use the average; otherwise a 5-minute and 10-

minute count should be adequate to obtain a reasonable value for the background radiation.

PART 3: PENETRATING POWER OF BETA RAYS

This part of the experiment should present no special problems to the student. Depending on your samples, you may want to change the distances between the sample and Geiger tube from those suggested in the experiment. If your students are familiar with logarithms, you may want to have them make a log plot of activity versus distance. Such a plot would yield a straighter line than the nonlogarithm plot.

PART 4: PENETRATING POWER OF GAMMA RAYS

This part of the experiment is also very straightforward. Tell students not to move the gamma source once they have positioned it at the start of the experiment.

You may also want your students to make a logarithm plot of activity versus number of lead plates. You should also discuss extrapolation of their curve or straight line so that they may obtain the number of lead plates necessary to bring the counts per minute of their gamma source down to zero.

Note: The lead plates we use are approximately 8 cm long, 2 cm wide, and 0.2 cm thick.

Some Possible Answers to the Questions

(Please note that the following are suggested answers to the questions, based on our experience with this course over the years; however, they are not the only possible answers.)

1. The Geiger tube does not count all the radiation being emitted from the sample, because the sample emits radiation in all directions, and the Geiger tube counts only those radiations that strike it. Also the Geiger tube suffers from a problem called resolving time. This means that the Geiger tube can't react fast enough to the constant stream of radiation hitting it, and so it misses some radiation.

2. Radiation is a spontaneous phenomenon, and although a sample emits about the same amount of radiation per unit of time, it is not an absolute amount, but instead varies about some average.

3. Another reason that the count of the radioactivity of the sample drops off as the sample is moved an increasing distance from the tube is that, as the Geiger tube moves farther away, more and more radiation misses it. It's like a flashlight beam that spreads out as you aim it farther and farther away.

EXPERIMENT 21

ORGANIC COMPOUNDS: A LOOK AT SOME DIFFERENT KINDS

Time About 2 hours

Materials For a section of 24 students working in groups of two: pentane (100 mL); ethyl alcohol (100 mL); diethyl ether (100 mL); cinnamaldehyde (100 mL); acetone (100 mL); 5-percent acetic acid (100 mL); amyl acetate (100 mL); methylamine (100 mL); toluene (100 mL); cyclopentane (100 mL); cyclohexane (100 mL); salicylic acid (100 g); methyl alcohol (100 mL); concentrated sulfuric acid (100 mL).

Discussion of the Experiment

This experiment enables the student to:

1. Examine some organic compounds and gain some familiarity with them.

2. Perform some simple reactions with organic compounds, in order to see how they behave.

Although the procedures in this experiment are simple, you must impress on the students the need for safety. Breathing and igniting small amounts of organic compounds are not dangerous if students follow proper safety rules and use good techniques. If you have any doubts about the capabilities or mature attitudes of your students, perform Part 2 of the experiment – Hydrocarbons and Flames – as a demonstration. Also, if any of your students have respiratory diseases or congestions, do not let them perform Part 1 of the experiment – Odor Characteristics of Organic Compounds.

PART 1: ODOR CHARACTERISTICS OF ORGANIC COMPOUNDS

Give each group of students a set of test tubes containing 1 to 2 mL of the organic compounds to be examined. Label these test tubes by number and name, except for the unknowns, which should be given a number only. Let the unknowns be members of the same classes as those tested, but let them be different compounds.

Most students do not know the proper technique for smelling a compound. It is important that you show them this technique before they perform the experiment.

Caution: Methylamine has an intense odor and should be handled with care.

PART 2: HYDROCARBONS AND FLAMES

When the students are performing this part of the experiment, make sure that there are no large containers of flammable materials in the laboratory. Also make sure that students remove all papers, books, and other materials from their work areas. This part of the experiment should be performed in a fume hood.

PART 3: A SIMPLE ORGANIC REACTION

In this part of the experiment students perform an esterification reaction, with methyl salicylate, commonly known as oil of wintergreen, as the product. There is no attempt to isolate the product; students simply prove its existence by its odor. The procedure and the amounts we suggest should yield sufficient methyl salicylate so that students may detect it by its smell.

Caution: Tell the students that when they heat their reaction mixture in the water bath, the methyl alcohol will boil. They should not allow their reaction mixture to boil so vigorously that it will boil over the sides of the test tube. If necessary they should stop heating the water bath.

Some Possible Answers to the Questions

(Please note that the following are suggested answers to the questions, based on our experience with this course over the years; however, they are not the only possible answers.)

1. Students will answer yes and no for this question. Ask them to support their answers.

2. Some of the combustion products are C, CO_2, H_2O.

3. The reaction is

$$C_6H_4(OH)-\overset{\displaystyle O}{\overset{\|}{C}}-OH + CH_3-OH \xrightarrow{H^+} C_6H_4(OH)-\overset{\displaystyle O}{\overset{\|}{C}}-OCH_3 + H_2O$$

The general reaction is

$$R-\underset{\underset{O}{\|}}{C}-OH + R'-OH \longrightarrow R\underset{\underset{O}{\|}}{C}-OR' + H_2O$$

4. The word ether was probably given to these compounds because of their high volatility, especially diethyl ether.

EXPERIMENT 22

CARBOHYDRATES, LIPIDS, AND PROTEINS

Time About 2 hours

Materials (For a section of 24 students working in groups of two): 2% starch (1 liter); diabetic urine (50 g of glucose per liter of distilled water, add yellow food coloring); 2% glucose solution (1 liter); iodine solution (dissolve 4 g of potassium iodide in 100 mL of distilled water, then add 2 g of iodine); Benedict's solution (1 liter); glycerol (100 mL); cottonseed oil (100 mL); olive oil (100 mL); 1% biuret (1 liter); 2% albumin (1 liter); milk (1 pint); 10% NaOH (1 liter); 0.5% $CuSO_4$ (1 liter); white bread (6 slices); butter (1 stick); brown paper.

Discussion of the Experiment

The main objective of this experiment is to have the student learn to test for certain carbohydrates, lipids, and proteins. The experiment is divided into three parts:

1. Testing for carbohydrates (glucose and starch).
2. Testing for lipids.
3. Testing for proteins.

The student should not encounter many problems when performing this experiment. The results are best when the solutions are fresh. If diabetic urine for Part 1 is not available, you can prepare diabetic urine by adding 50 g of glucose to a liter of water and dropping in yellow food coloring. Be sure that the students wear their safety glasses throughout the experiment.

Some Possible Answers to the Questions

1. Glucose solution and diabetic urine were positive results. Glucose should not appear in normal urine. Therefore the results should be different for normal and diabetic urine.
2. Spaghetti and paper should be positive. Apple juice should be negative.
3. In the acrolein test, fats undergo hydrolysis. By action of dehydrating agents and high temperatures, glycerol, one of the hydrolysis products, is converted to acrolein (which has a pungent odor).
4. Positive results for 1% biuret, 2% albumin, and milk should be noted.

EXPERIMENT 23

DETERMINATION OF THE AMOUNT OF IRON IN WATER

Time About 2 hours

Materials (For a section of 24 students working in groups of two): colorimeters (12, or can get by with fewer if necessary); standard iron solution (10 mg/liter) prepared according to the directions to follow (1 liter); unknown iron solution(s) prepared according to the directions to follow (1 liter of each); FerroVer ® powder pillows* (60 pillows, if each group does only one unknown); 100-mL beakers (60); 25-mL graduated cylinders (12).

PREPARATION OF SOLUTIONS

1. Preparation of stock iron(II) nitrate solution (100 mg/liter Fe^{+2}): Dissolve 0.516 gram of $Fe(NO_3)_2 \cdot 6H_2O$ in enough water to make exactly one liter of solution.

2. Preparation of standard iron solution (10 mg/liter Fe^{+2}): Dilute exactly 100 mL of the stock iron solution with enough water to make one liter of solution.

3. Preparation of unknown iron solutions: Prepare exactly one liter of each of the following unknown iron solutions by diluting the specific amounts of stock iron solution with enough water to make one liter of solution. (See table that follows.)

Unknown no.	Concentration of Fe^{+2}	Milliliters of stock Fe^{+2} solution (100 mg/liter)
1	1.0 mg/liter	10 mL
2	1.5 mg/liter	15 mL
3	2.0 mg/liter	20 mL
4	3.0 mg/liter	30 mL

* FerroVer ® Powder Pillows may be obtained from the Hach Chemical Company, Ames, Iowa.

Discussion of the Experiment

From this experiment the student should learn to:

1. Operate a colorimeter.

2. Explain the principle behind the colorimetric determination of iron.

3. Plot a graph of absorbance versus concentration for the iron standards and use this graph to find the concentration of the unknown iron solutions.

We recommend that you review the operation of the colorimeter with your class before starting the experiment. You may also want to review the calculations for dilutions of solutions, so that the students know how the numbers in Step 3 of Procedure 2 were calculated.

The experiment is written so that you can decide how many unknowns you want each group of students to analyze. Each group will need four powder pillows plus an additional powder pillow for each unknown to be analyzed. These powder pillows, which are inexpensive, are easily obtained from Hach Chemical Company in Ames, Iowa.

Although we suggest using graduated cylinders to measure the 25 mL of each sample, you may want to substitute 25-mL pipettes if they are available. We find that graduated cylinders are adequate for our purposes.

Some Possible Answers to the Questions

(Please note that the following are suggested answers to the questions, based on our experience with this course over the years; however, they are not the only possible answers.)

1. The iron must be in the iron(II) state, Fe^{+2}.

2. High concentrations of iron in potable water supplies can interfere with laundering operations, impart stains to plumbing fixtures, and support the growth of bacteria.

3. You would have to dilute the sample quantitatively.

EXPERIMENT 24

DETERMINATION OF THE DISSOLVED-OXYGEN CONTENT OF WATER

Time About 2 hours

Materials (For a section of 24 students working in groups of two): 300-mL BOD bottles (12) or Erlenmeyer flasks (you may use 300-mL or larger bottles instead if BOD bottles are not available); graduated cylinders (12 100-mL size and 12 10-mL size); burettes (12); manganese sulfate solution (1 liter) prepared according to directions to follow; alkali-iodide-azide reagent (1 liter) prepared according to directions to follow; concentrated sulfuric acid (500 mL); 0.025N sodium thiosulfate (1 liter) prepared according to directions to follow; starch solution (1 liter) prepared according to directions to follow; 500-mL Erlenmeyer flasks (12); plunger pipettes (2), calibrated to deliver 2 mL (standard 2-mL pipettes may be used if plunger pipettes are not available); water samples (you may use tap water, or lake, or river water, in sufficient quantity so that each group has at least 300 mL for each trial run).

PREPARATION OF SOLUTIONS

1. Manganese sulfate solution:
   Dissolve 480 g of $MnSO_4 \cdot 4H_2O$ in sufficient water to make one liter of solution. (Alternatively, you may use 400 g of $MnSO_4 \cdot 2H_2O$, or 364 g of $MnSO_4 \cdot H_2O$.)

2. Alkali-iodide-azide reagent:
   Dissolve 500 g of sodium hydroxide and 135 g of sodium iodide in enough water to prepare 1 liter of solution. To this solution add 10 g of sodium azide (NaN_3) dissolved in 40 mL of water.

   (Note: 700 g of KOH may be substituted for the 500 g of NaOH, and 150 g of KI many be substituted for the 135 g of NaI.)

3. Starch solution:
   Add about 5 g of soluble starch to 500 mL of boiling water. Allow this mixture to boil for about ten minutes with stirring, then dilute to one liter. Add about 1.3 g of salicylic acid to this starch solution to preserve it. (If you have time to allow the starch solution to stand overnight, you can then decant the clear supernatant from the rest of the solution. You may use this clear supernatant part for best results.)

4. Standard 0.025N sodium thiosulfate solution:
   Dissolve 6.205 g of $Na_2S_2O_3 \cdot 5H_2O$ in freshly boiled distilled or deionized water and dilute to one liter. (Actually this solution should be standardized against a primary standard such as potassium dichromate, but for our purposes this need not be done.)

Discussion of the Experiment

From this experiment the student should learn to:

1. Discuss the importance of dissolved oxygen in water.
2. Test for the amount of dissolved oxygen in a sample of water.
3. Perform an oxidation-reduction titration.

Although this experiment requires a number of solutions to be prepared in advance by the stockroom, most students find it an extremely interesting experiment to run. The reactions involved are as follows:

$$Mn^{+2} + 2OH^{-1} \longrightarrow \underset{\text{White precipitate}}{Mn(OH)_2}$$

(This reaction occurs if no oxygen is present in the sample.)

$$Mn^{+2} + 2OH^{-1} + \frac{1}{2}O_2 \longrightarrow MnO_2 + H_2O$$

$$Mn(OH)_2 + \frac{1}{2}O_2 \longrightarrow \underset{\text{Brown precipitate}}{MnO_2} + H_2O$$

$$MnO_2 + 2I^{-1} + 4H^{+1} \longrightarrow Mn^{+2} + I_2 + 2H_2O$$

The amount of iodine produced depends on the amount of oxygen originally present in the sample. The iodine is titrated with sodium thiosulfate according to the following equation.

$$2S_2O_3^{-2} + I_2 \longrightarrow S_4O_6^{-2} + 2I^{-1}$$

Although we usually don't bother with the equations in this course, a simplified set of equations showing the chemical reactions are presented in the experiment. We find that this is more than sufficient for the student.

When performing the experiment, we find that it is best to use BOD bottles that have a 300-mL capacity. These can be purchased from any scientific supply house. Otherwise, you may use other appropriate glassware. We also find it advisable to use plunger pipettes in order not to aerate the sample. However, you may use standard pipettes.

In titrating the last remaining traces of iodine, you may want to explain the function of the starch to your students.

All in all, this experiment is pretty easy for the students to run. However, make sure that the students wear their glasses during the entire experiment, and handle the reagents carefully.

Some Possible Answers to the Questions

(Please note that the following are suggested answers to the questions, based on our experience with this course over the years; however, they are not the only possible answers.)

1. Needed to maintain fish and aerobic microorganisms

2. $Mn^{+2} + O_2 \rightarrow MnO_2$

   Mn^{+2} Oxidized, O_2 Reduced

   $MnO_2 + I^{-1} \rightarrow Mn^{+2} + I_2$

   MnO_2 Reduced, I^{-1} Oxidized

   $I_2 + S_2O_3^{-2} \rightarrow I^{-1} + S_4O_6^{-2}$

   I_2 Reduced, $S_2O_3^{-2}$ Oxidized

EXPERIMENT 25

CHEMICAL KINETICS - RATES OF REACTION

Time About 2 hours

Materials (For a section of 24 students working in groups of two: 0.1M sodium oxalate (1 liter); 0.1M potassium permanganate (1 liter); 6M sulfuric acid (1 liter); 0.2M potassium iodate (labeled solution A, 2 liters); 0.1M acidified sodium sulfite solution with starch (labeled solution B, 2 liters); 0.1M copper (II) sulfate (1 liter); deionized or distilled water, 400-mL beakers (12); graduated cylinders (10-mL, 25-mL, and 100-mL, 12 of each) or pipettes (various sizes); thermometers (12).

PREPARATION OF SOLUTIONS

1. Sodium oxalate solution (0.1M)

   Dissolve 13.4 g of $Na_2C_2O_4$ in enough water to prepare 1 liter of solution.

2. Potassium permanganate solution (0.1M)

   Dissolve 15.8 g of $KMnO_4$ in enough water to prepare 1 liter of solution.

3. Potassium iodate solution (0.2M) [Solution A]

   Dissolve 8.6 g of KIO_3 in enough water to prepare 2 liters of solution.

4. Starch solution:

   Add about 10 g of soluble starch to 500 mL of boiling water. Allow this mixture to boil for about ten minutes with stirring. (If you have time, allow the starch solution to stand overnight. You can then decant the clear supernatant from the rest of the solution. You may use this clear supernatant part for best results.)

5. Acidified sodium sulfite solution with starch (0.1M) [Solution B]

Into a 2-liter volumetric flask dissolve 2.6 g of Na_2SO_3 and 20 mL of 6M H_2SO_4 in about 500 mL of water. To this add the 500 mL of starch solution that was just prepared (or all of the clear supernatant, if that's what you're using) and stir. Add sufficient water to bring the volume up to 2 liters.

6. Copper (II) sulfate solution (0.1M)

Dissolve 25 g of $CuSO_4 \cdot 5H_2O$ in enough water to prepare 1 liter of solution.

Discussion of the Experiment

The main purpose of this experiment is to impress on the minds of the students that not all reactions occur simultaneously. Also, the students will learn that even the same reaction can occur at different rates depending on the reaction variables of concentration, temperature, and catalyst.

Students usually enjoy performing this experiment because it has a kind of "magic" quality to it. However, be sure to point out to them that they are performing a carefully controlled experiment and that it is important for them to label all of their solutions carefully and perform their measurements exactly.

Some Possible Answers to the Questions

(Please note that the following are suggested answers to the questions, based on our experience with this course over the years; however, they are not the only possible answers.)

1. The iodine standard clock reaction will occur more quickly than the sodium oxalate/potassium permanganate reaction.

2. In general, the reaction rate will be faster for each of the solutions prepared in Part 2A than for the standard run.

3. (a) The reaction rate should be slowed by half for each 10°C drop in temperature.

   (b) The reaction rate should double for each 10°C rise in temperature.

4. The reaction rate should speed up with the addition of the copper (II) sulfate catalyst.

PART V

TRANSPARENCIES

PART V

TRANSPARENCIES

During the many years that we have been teaching introductory chemistry courses, we have found that we need overhead transparencies of certain key tables and figures in the textbook. It saves an instructor time and effort (and helps the student grasp ideas more quickly) when the instructor can point to something on a large reproduction of an illustration and say, "Let's go over this one step at a time." To save ourselves the time it took to write everything on the board, we cut certain illustrations out of the textbook, had them reproduced in a larger size, made Thermofax or acetate copies, and used these as overhead transparencies in class. It worked well. We were able to cover material in greater detail, and the students seemed to like this method of teaching.

And so we have decided to include a set of 52 overhead transparencies with our basic chemistry package. We chose 28 tables and 23 figures from the text (some required more than a single transparency) and had them blown up and printed. We hope that every instructor who adopts the text will find, among the transparencies, those items that he or she considers most useful. We should greatly appreciate hearing reactions from users of the text. We wish you good teaching!

TRANSPARENCY TITLES

Table or Figure Number	Description
Figure 2.5	A comparison of Fahrenheit and Celsius scales
Table 2.3	Basic SI (metric) units
Table 2.4	Metric prefixes
Table 2.5	Conversion of units
Figure 4.1	A Crookes or cathode-ray tube
Figure 4.4	The Thomson model of the atom
Figure 4.5	Prediction based on Thomson model
Figure 4.6	Rutherford's experiment
Table 4.1	The particles in the atom
Table 4.2	Naturally occurring isotopes of the first 15 elements
Figure 5.2	A spectroscope
Figure 5.3	A continuous spectrum and a line spectrum
Figure 5.4	The electromagnetic spectrum
Figure 5.10	Energy levels and sublevels
Figure 5.11	The shapes of the s, p_x, p_y, and p_z orbitals
Figure 5.12	The relationship of the 1*s*, 2*s*, and 3*s* orbitals
Figure 5.16	How energy levels overlap
Figure 5.17	Sublevel filling order
Table 5.1	Theoretical maximum energy level capacity

Table or Figure Number	Description
Table 5.2 (part 1)	Ground-state configurations
Table 5.2 (part 2)	Ground-state configurations
Figure 6.3	Newlands's law of octaves
Figure 6.8	The decrease of atomic radii within a period
Figure 6.9	The decrease of ionization potential within a group
Table 6.5	Summary of periodic trends
Figure 7.12	Three-dimensional models of HCl, H_2O, and CH_4
Table 8.2	Charges of ions
Table 8.3	Charges of ions
Table 8.4	Common names of some chemical compounds
Table 10.1	The activity series
Figure 12.2	Reaction diagrams
Table 12.1	Specific heat of some substances at 25°C
Table 12.2	Heats of formation at 25°C and 1 atm pressure
Table 13.1	Pressure of water vapor at various temperatures
Figure 14.8	Phase change diagram of H_2O
Figure 14.14	Six types of crystals
Table 14.2	Some properties of crystalline solids
Table 15.2	Solubility of various substances in water at 0°C and 50°C
Table 16.1	Some common Arrhenius acids
Table 16.2	Some common Brønsted-Lowry bases
Table 16.3	Some common Arrhenius bases
Table 16.4	Some common Brønsted-Lowry bases
Table 16.8	The relationship between hydrogen-ion concentration and hydroxide-ion concentration
Figure 17.1	Endothermic and exothermic reaction diagrams
Figure 17.2	Activation energy
Figure 17.3	Effect of a catalyst on activation energy
Table 17.1	Ionization constants of weak acids and bases at 25°C
Table 18.1	Summary of nuclear particles
Table 19.1	The first ten members of the alkane series
Table 19.3	Names of some alkyl groups
Table 19.4	The alkene and alkyne series
Table 20.1	Some classes of organic compounds